Christof Stieger

Implantable Hearing Devices

Christof Stieger

Implantable Hearing Devices

Developement and Evaluation of two concepts

Südwestdeutscher Verlag für Hochschulschriften

Impressum/Imprint (nur für Deutschland/ only for Germany)
Bibliografische Information der Deutschen Nationalbibliothek: Die Deutsche Nationalbibliothek verzeichnet diese Publikation in der Deutschen Nationalbibliografie; detaillierte bibliografische Daten sind im Internet über http://dnb.d-nb.de abrufbar.
Alle in diesem Buch genannten Marken und Produktnamen unterliegen warenzeichen-, marken- oder patentrechtlichem Schutz bzw. sind Warenzeichen oder eingetragene Warenzeichen der jeweiligen Inhaber. Die Wiedergabe von Marken, Produktnamen, Gebrauchsnamen, Handelsnamen, Warenbezeichnungen u.s.w. in diesem Werk berechtigt auch ohne besondere Kennzeichnung nicht zu der Annahme, dass solche Namen im Sinne der Warenzeichen- und Markenschutzgesetzgebung als frei zu betrachten wären und daher von jedermann benutzt werden dürften.

Verlag: Südwestdeutscher Verlag für Hochschulschriften Aktiengesellschaft & Co. KG
Dudweiler Landstr. 99, 66123 Saarbrücken, Deutschland
Telefon +49 681 37 20 271-1, Telefax +49 681 37 20 271-0, Email: info@svh-verlag.de
Zugl.: Neuchâtel, UniNE, Diss., 2006

Herstellung in Deutschland:
Schaltungsdienst Lange o.H.G., Berlin
Books on Demand GmbH, Norderstedt
Reha GmbH, Saarbrücken
Amazon Distribution GmbH, Leipzig
ISBN: 978-3-8381-0584-0

Imprint (only for USA, GB)
Bibliographic information published by the Deutsche Nationalbibliothek: The Deutsche Nationalbibliothek lists this publication in the Deutsche Nationalbibliografie; detailed bibliographic data are available in the Internet at http://dnb.d-nb.de.
Any brand names and product names mentioned in this book are subject to trademark, brand or patent protection and are trademarks or registered trademarks of their respective holders. The use of brand names, product names, common names, trade names, product descriptions etc. even without a particular marking in this works is in no way to be construed to mean that such names may be regarded as unrestricted in respect of trademark and brand protection legislation and could thus be used by anyone.

Publisher:
Südwestdeutscher Verlag für Hochschulschriften Aktiengesellschaft & Co. KG
Dudweiler Landstr. 99, 66123 Saarbrücken, Germany
Phone +49 681 37 20 271-1, Fax +49 681 37 20 271-0, Email: info@svh-verlag.de

Copyright © 2009 by the author and Südwestdeutscher Verlag für Hochschulschriften
Aktiengesellschaft & Co. KG and licensors
All rights reserved. Saarbrücken 2009

Printed in the U.S.A.
Printed in the U.K. by (see last page)
ISBN: 978-3-8381-0584-0

Contents

CHAPTER 1: INTRODUCTION 15
    1.1   PURPOSE OF THIS WORK 16
    1.2   ANATOMY AND PHYSIOLOGY OF THE EAR 16
    1.3   HEARING DISORDERS..20
    1.4   THERAPIES 22
    1.5   REFERENCES 28

CHAPTER 2: COMPUTER ASSISTED OPTIMIZATION OF AN ELECTROMAGNETIC TRANSDUCER DESIGN FOR IMPLANTABLE HEARING AIDS...31
    2.1   ABSTRACT...32
    2.2   INTRODUCTION..32
    2.3   A SIMPLE ELECTROMAGNETIC TRANSDUCER 34
    2.4   FACTORS LIMITING THE DESIGN PARAMETERS 36
    2.5   MATERIAL AND METHODS 38
    2.6   RESULTS 41
    2.7   DISCUSSION..44
    2.8   SUMMARY 46
    2.9   REFERENCES 47

CHAPTER 3: ANATOMICAL STUDY OF THE HUMAN MIDDLE EAR FOR THE DESIGN OF IMPLANTABLE HEARING AIDS 49
    3.1   ABSTRACT...50
    3.2   INTRODUCTION..50
    3.3   MATERIAL AND METHOD 52
    3.4   RESULTS 55
    3.5   DISCUSSION..60
    3.6   CONCLUSIONS 62

3.7 REFERENCES 62

CHAPTER 4: HUMAN TEMPORAL BONES VERSUS MECHANICAL MODEL TO EVALUATE THREE MIDDLE EAR TRANSDUCERS 65

    4.1 ABSTRACT 66
    4.2 INTRODUCTION 66
    4.3 METHODS 68
    4.4 RESULTS 74
    4.5 DISCUSSION 80
    4.6 CONCLUSIONS 82
    4.7 REFERENCES 83

CHAPTER 5: A NOVEL IMPLANTABLE HEARING SYSTEM WITH DIRECT ACOUSTICAL COCHLEAR STIMULATION (DACS) 87

    5.1 ABSTRACT 88
    5.2 INTRODUCTION 89
    5.3 METHODS 90
    5.4 RESULTS 97
    5.5 DISCUSSION 104
    5.6 REFERENCES 106

CHAPTER 6: CONCLUSION AND OUTLOOK 111

    6.1 CONTACTLESS TRANSDUCER (CLT) 112
    6.2 DIRECT ACOUSTICAL COCHLEAR STIMULATION (DACS) 113
    6.3 OUTLOOK 114

Abstract

For the treatment of hearing impairments, various types of hearing devices and surgical methods exist. However, these methods are not adequate for all types of hearing losses and patients' needs. Depending on the design, some of the inherent shortcomings could be reduced by implantable hearing systems. In implantable hearing systems, the output is generated by an implantable transducer, i.e. the equivalent of the loudspeaker of a conventional hearing aid.

Two principles of implantable hearing systems have been evaluated and developed in this thesis. One of the presented transducers (i.e. contactless transducer, CLT) is intended for the rehabilitation of sensorineural hearing loss and couples to ossicles of the middle ear. It consists of a coil and magnet, both implantable in the middle ear in a minimally invasive manner. The second (i.e. DACS) is intended for the rehabilitation of severe combined perceptive and conductive hearing loss. It couples directly to the inner ear fluid by means of an amplified stapes prosthesis.

Computer tomography (CT) scans of isolated human heads were used to generate morphometrical data. These data defined transducer geometry. Computer simulation were used to optimize the CLT. Both transducers have been measured and evaluated using Laser Doppler vibrometry (LDV). For the CLT those measurements where performed on a mechanical middle ear model and on human temporal bones. For the DACS isolated human heads were measured and because of favourable results the system was implanted in three patients. Audiological tests were performed pre- and post-operatively for the DACS.

Optimization showed that CLT can be constructed to generate maximal output and to facilitate the implantation procedure. Mounting parameters have an important influence on the output of the CLT. The overall output is high for low frequencies and high frequencies but it was insufficient in the middle frequency range which is particularly important for speech intelligibility. The primary goal of a minimally invasive implantable transducer was reached with the CLT, but an overall high output was not obtained. Therefore, this concept was not pursued.

DACS is an abbreviation for Direct Acoustical Cochlear Stimulation. It was implanted in three patients. The newly developed surgical procedure was applied without any unexpected troubles. The generated output of all transducers at 1 mW was more than 125 dB SPL for the entire audiological relevant frequency range. All patients showed significantly improved hearing with the implanted DACS. Such results could hardly have been reached with a conventional hearing aid or surgical intervention alone. Because of the favourable first results, further development or the DACS is undertaken.

Kurzfassung

Zur Behandlung von Schwerhörigkeiten stehen heute verschiedene Hörsysteme, sowie die Mittelohrchirugie zur Verfügung. Trotzdem kann nicht allen Patienten zufriedenstellend geholfen werden. Eine Verbesserung wäre bei gewissen Patienten mit implantierbaren Hörsystemen zu erreichen. Implantierbare Hörsysteme haben im Gegensatz zu konventionellen Hörgeräten keinen Lautsprecher, sondern einen implantierten Transducer.

In der vorliegenden Arbeit werden zwei speziell entwickelte implantierbare Transducer vorgestellt. Beim ersten Transducer handelt es sich um ein minimal invasiv implantierbares Design CLT (Contactless Transducer). Dieses ist zur Behandlung von Schallempfindungs-Schwerhörigkeiten ausgelegt worden. Der CLT besteht aus einer Spule, welche auf einem der Gehörknöchelchen befestigt wird und einem Permanentmagneten, welcher an der Wand in der Paukenhöhle fixiert wird. Der zweite Transducer DACS (Direct Acoustical Cochlea Stimulation) wurde für die Behandlung von kombinierten Schwerhörigkeiten konzipiert. Im Gegensatz zum CLT wird der DACS Transducer mittels einer sogenannten Stapesprothese direkt an die Innenohrflüssigkeit gekoppelt.

Mit Computertomographie (CT) wurden anatomische Messungen durchgeführt, welche für die Konstruktion der Transducer nötig waren. Dazu wurden anatomische Ganzkopfpräparate verwendet. Die Optimierung des CLT Transducers erfolgte mittels Computersimulation. Beide Transducer wurden mit Laser Doppler Vibrometrie gemessen und evaluiert. Für Messungen des CLT's wurden frische Felsenbeine sowie ein mechanisches Mittelohrmodel verwendet. Der DACS Transducer wurde zuerst an anatomischen Ganzkopfpräparaten gemessen und aufgrund der günstigen Resultate bei drei Patienten implantiert. Bei diesen Patienten wurden prä- und postoperative audiologische Tests durchgeführt.

Der CLT kann so optimiert werden, dass er maximale Ausgangsverstärkung generiert oder tolerantes Verhalten bezüglich Positionierung aufweist. Für tiefe und hohe Frequenzen wurden relativ grosse Verstärkungen erzielt. Im mittleren Frequenzbereich, welcher besonders wichtig für das Sprachverständnis ist, blieb die Ausgangsverstärkung jedoch ungenügend. Das primäre Ziel der Entwicklung eines minimal invasiv implantierbaren Transducers konnte zwar erreicht werden. Auf Grund der ungenügenden Verstärkung im mittleren Frequenzbereich, wurde die Entwicklung dieses Transducers jedoch nicht mehr vorangetrieben.

Der DACS Transducer wurde in drei Patienten implantiert. Dafür wurde ein neuer, sogenannter retrokanalärer, chirurgischer Zugang zum Mittelohr entwickelt. Für den Transducer wurde breitbandig ein äquivalenter Ausgangsschalldruck von über 125 dB SPL erreicht. Bei allen Patienten war das Hörvermögen postoperativ erheblich besser als präoperativ. Solche Ergebnisse wären durch die Chirugie allein oder durch Anpassung eines konventionellen Hörgerätes nicht möglich gewesen. Da die Resultate dieses Konzepts sehr vielversprechend sind, wird die Entwicklung weiterverfolgt.

Résumé

Le traitement de la surdité est basé soit sur l'appareillage acoustique soit sur la microchirugie otologique. Ces méthodes ne sont toutefois pas efficaces pour tous les types de défauts auditives. Dans certains cas une amélioration est possible par des systèmes auditives implantables. Ces systèmes possèdent tous un transducteur implantable qui correspond au haut-parleur des prothèses acoustiques conventionnelles.

Dans cette thèse deux principes de transducteur implantable ont été développés et évalués. Le premier transducteur (CLT) est construit pour la réhabilitation des default auditif de la perception (neurosensorielle) et se greffe dans l'oreille moyenne. Il est composé d'une bobine fixée au marteau et d'un aimant permanent attaché au bord de la cavité de l'oreille moyenne. Le deuxième transducteur (DACS) a été développé pour la réhabilitation des surdités mixed ayant, en plus une composante de transmission. Dans ce système les vibrations sonores sont transmises directement au liquide de l'oreille interne à l'aide d'un prothèse de stapedectomie qui est relié au amplificateur électromécanique (transducteur).

La géométrie des transducteurs a été élaborée à partir d'images tomographiques de crânes. Leur caractérisation a été étudier au moyen d'un Laser Doppler Vibromètre (LDV), sur un modèle mécanique de grandeur nature ainsi que sur des os temporaux pour le CLT et sur des crânes pour le DACS. En plus, la performance du DACS a, pu être confirmée lors de l'opération de trois patients. Pour son implantation, un nouvel abord chirurgical de l'oreille moyenne a été développé. Finalement, le CLT a été optimisé à l'aide de simulations numériques.

Les simulations numériques ont montré que le CLT peut, soit être optimisé au niveau de l'amplification, soit au niveau de la procédure d'alignement de la bobine et de l'aimant. ¨Les mesures sur le modèle mécanique ont permis de mettre en évidence les paramètres ayant une grande influence sur l'amplification. Pour les hautes et basses fréquences les résultats sont encourageants, mais dans les fréquences moyennes c'est-à-dire dans les fréquences du langage, l'amplification reste faible. Le but principal qui consiste à développer un transducteur implantable avec une chirurgie invasive minimale est atteint. Mais les performances insuffisantes de l'amplification ont conduit à l'abandon du développement de ce transducteur.

Le système DACS a été implanté à trois patients. L'amplification mesurée pendant l'opération était de plus de 125 dB SPL sur toutes les fréquences. L'audition des patients implanté est par la suite considérablement améliorée grâce au système DACS. L'amélioration était supérieur à celle obtenu par seulement l'appareillage acoustique ou par l'intervention chirugicale de stapedectomy. En raison de ces résultats favorables le développement du DACS est poursuivi.

List of Acronyms

Acronym	Full Name
B	Magnetic flux density
BAHA	Bone anchored hearing aid
BMEC	Border of middle ear cavity
BTE	Behind the ear
CC	Crimp connection
CI	Cochlear implant
CIC	Complete in the canal
CLT	Minimally invasive implantable transducer / Contactless transducer
CO	Coil
CR	Coupling rod
CT	Computer tomography
d_i	Inner diameter of the coil
d_m	Diameter of permanent magnet
d_o	Outer diameter of coil
dB	Dezibel
DP	Degree of Pneumatization
DRT	Direct rod transducer
EMBEC	European medical and biological engineering conference
ENT	Ear nose throat
FDA	Food and drug association of the United States
FMT	Floating mass transducer

Acronym	Full Name
H	Magnetic field
h	Height of coil
h_m	Heigth of permanent magnet
HL	Hearing level
Hz	Hertz
I	Current in the coil
IHC	Inner hair cell
IHS	Implantable hearing system (Type 0-III)
IN	Incus
ITC	In the canal
ITE	In the ear
j	Current density
L	Length of the wire of the coil
LDV	Laser Doppler vibrometry
LIP	Ligamentum incudis posterior
LMA	Ligamentum mallei anterior
MA	Malleus
MM	Malleus mallei (handle of the malleus)
MES	Middle ear surgery
N	Number of turns of the coil
OHC	Outer hair cell
OW	Oval window
PM	Permanent magnet
PP	Percutaneous plug
q^2	Cross section of a single wire
RF	Radiofrequency
RP	Reference point

Acronym	Full Name
RW	Round window
SP	Stapes prosthesis
SPL	Sound pressure level
ST	Stapes
SUVA	Schweizerische Unfallversicherungsanstalt
T	Transducer
TICA	Totally implantable cochlear amplifier
z	Air gap between the coil and magnet
η	Filling factor of the coil
μ_r	Relative permeability
ρ	Radial displacement between coil and magnet

Chapters overview

This thesis contains following chapters:

Chapter 1 :"Introduction" defines the purpose of this dissertation, provides a general introduction into the anatomy, physiology, and pathology (diseases) of hearing and current possibilities for rehabilitation and consequent motivation for the utilization of implantable hearing aids.

Chapter 2 : "Computer assisted optimization of an electromagnetic transducer design for implantable hearing aids" introduces the minimally invasive, implantable middle ear transducer (CLT). It consists of a coil and a magnet to be implanted in the middle ear cavity. In order to optimize the dimensions of the coil, a set of four simulations for different geometries were performed and experimentally verified. As a result, the coil can be optimized either to maximize output levels or to be tolerant of radial displacements between coil and magnet.
Stieger C, Wackerlin D, Bernhard H, Stahel A, Kompis M, Hausler R, Burger E. Computer assisted optimization of an electromagnetic transducer design for implantable hearing aids. Computers in Biology and Medicine 2004;34 (2):141-52.

Chapter 3: "Anatomical study of the human middle ear for the design of implantable hearing aids" generates a set of morphometric data which is mandatory for the design of the CLT and the DACS. The data is generated using computer tomography (CT) scans of human heads *post mortem*. Data were statistically examined on different factors.
Stieger C, Djeric D, Kompis M, Remonda L, Hausler R. Anatomical study of the human middle ear for the design of implantable hearing aids. Auris Nasus Larynx 2006;33 (4):375-80.

Chapter 4: "Human temporal bones versus mechanical model to evaluate three middle ear transducers" consists of two parts. First, a mechanical middle ear model, used to characterize the contactless transducer (CLT), is presented. When developing transducers, it is important to know the generated output at the stage of the cochlea. For conventional hearing aids, this is performed measuring the sound pressure level (SPL). As IHS transducers generate forces or displacements, appropriate models are necessary to evaluate the transducers. For IHS I, temporal bone models are often used for characterization. This chapter shows the advantages of a mechanical middle ear model.

Then, the output of three middle ear transducers for different mounting parameters is discussed. The mechanical middle ear model is used to compare the CLT with other middle ear transducers.

Stieger C, Bernhard H, Waeckerlin D, Kompis M, Burger J, Häusler R. Human temporal bones versus mechanical model to evaluate threemiddle ear transducers. JRRD 2007;44:407-16.

Chapter 5: "A novel implantable hearing system with direct acoustical cochlear stimulation (DACS)" presents the concept of the DACS system. The implantation procedure is described. Finally, results of the first clinical study are provided. They show that patients with severe combined hearing loss can be treated effectively with this new concept.

Hausler R, Stieger C, Bernhard H, M. Kompis. A Novel Implantable Hearing System with Direct Acoustic Cochlear Stimulation. Audiol Neurootol 2008;13:247-56.

Chapter 1: Introduction

The introduction defines the purpose of the dissertation, provides a general introduction into anatomy, physiology pathology (diseases) of hearing and today's possibilities for rehabilitation. Consequently, the motivation for the development of implantable hearing systems is given.

1.1 Purpose of this work

The purpose of this work is:

- To present two concepts of innovative implantable hearing systems:
  - The first system is a middle ear transducer which couples to the ossicles of the middle ear. Implantation is intended to be minimally invasive.
  - The second system (DACS) couples mechanically directly to the fluid of inner ear.
- To acquire morphometrical data for the middle ear cavity in order to outline maximal dimensions which are to be used in the design of transducers.
- To characterize, optimize and evaluate the CLT using simulations and measurements on a mechanical model as well as on temporal bones. The CLT should reach comparable or higher outputs compared to other middle ear transducers.
- To prove the concept of DACS in a clinical study.

1.2 Anatomy and physiology of the ear

Hearing refers to the ability to detect sound. It comprises not only the ear but also the auditory nerve and central auditory system including the auditory cortex in the temporal lobe. The ear acts as a sensor for acoustical stimulation by transforming sound into neural signals. These signals are transmitted by the auditory nerve and processed and analysed in the central auditory system. In this section, anatomy and physiology is limited to the ear, as, at the present time, the means of treatment of hearing impairments interact with the ear (except brainstem implants). Parts of this section are adapted from Kompis [1].

The human ear consists of three parts: external ear, middle ear and the inner ear. Figure 1.1 shows a cross section of the human ear.

Introduction

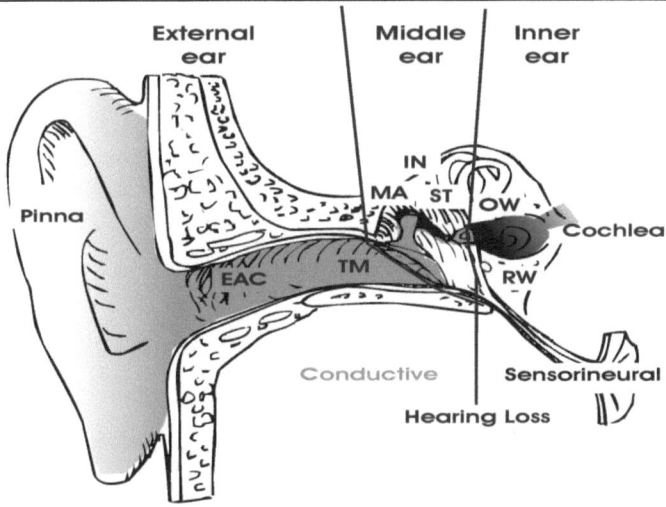

Figure 1.1 Cross section of the human ear. EAC: external auditory canal, TM: Tympanic membrane, MA: Malleus, IN: Incus, ST: stapes, OW: Oval Window, RW: Round Window (courtesy Kompis [1])

The frequency range of the audible field goes from 16 Hz –20'000 Hz and the dynamic is 100 dB.

1.2.1 External ear

The external ear consists on the auricle and the external auditory canal (Figure 1.1). Due to the structure of the auricle, sound is bundled and reaches the external auditory canal by multiple ways and with short time delays. The amplitude of sound is therefore amplified in some frequencies. As an approximation, the external the auditory canal acts as $\lambda/4$ resonator. Taking into account the average length of 23.4 mm (posterior wall) and 35.2 mm (anterior wall) [2] and a velocity of sound in the air of 343 m/sec, the resonance frequency is calculated between 2450 Hz and 3700 Hz (velocity divided by the fourfold length). As an example, the overall transfer function of the author's external ear is shown Figure 1.2.

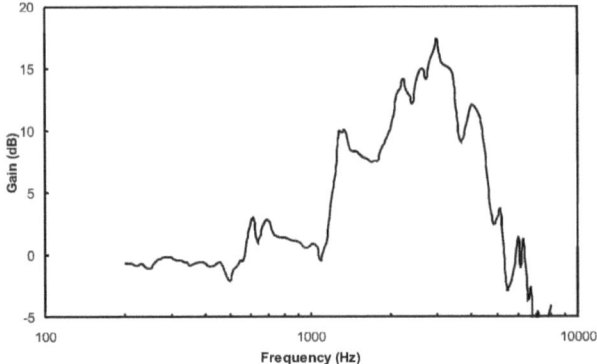

Figure 1.2 Measured transfer characteristic of the external ear of the author. A sweep of a sinus signal between 200 and 8000 Hz is applied. The gain represents the difference of sound pressure level measured by two microphones. One microphone is located 0.5 cm below the auricle and the second 2 mm in front of the tympanic membrane.

1.2.2 Middle ear

The middle ear consists of the tympanic membrane (TM) and tympanic cavity where three small ossicles are located.

The three ossicles, i.e. malleus (MA), incus (IN) and stapes (ST) are connected to each other. They transfer acoustic vibrations of the tympanic membrane to the inner ear fluid via the oval window. The function of the middle ear is the adaptation of the acoustical impedance Z (i.e. ratio between amplitude of pressure to velocity of the sound wave). The transmission of sound from the air into fluid is very poor.

$$Transmission\% = (1 - \frac{Z_{fluid} - Z_{air}}{Z_{fluid} + Z_{air}}) \cdot 100\% = 0.06\% \tag{1.1}$$

where:

$$Z_{air} = 413 \frac{kg}{s \cdot m^2} \tag{1.2}$$

$$Z_{fluid} = 482'030 \frac{kg}{s \cdot m^2} \tag{1.3}$$

Two middle ear mechanisms increase transmission of sound to liquid of the inner ear. First, the malleus and incus have different operative lengths. Therefore, by the law of lever, the force at the oval window is 1.3 times greater than at the tympanic membrane. Second, the ratio of the surface of the tympanic membrane (55 mm^2) and oval window (3.2 mm^2) is approximately 17 [2].

A combination of both effects amplifies the pressure at the oval window by 22.

Additionally, there are two muscles in the tympanic cavity, the stapedial muscle and the tensor tympani muscle. For acoustical signals above 70-100 dB HL (0 dB HL i.e. hearing level which refers to the physiological normal hearing), the stapedial muscle contracts with a short delay of approximately 100 ms. The stapedial muscle also contracts 100 ms before the onset of one's own voice. The reflex attenuates low frequencies below 2 kHz. Physiological explanations for this reflex are still under discussion; these include protection from loud acoustical events and protection from one's own voice, as well better speech intelligibility in noise[1].

1.2.3 Inner ear

The inner ear may be divided in two parts. The posterior part of the inner ear (vestibular labyrinth) is responsible for orientation in space; the anterior part (cochlea) acts as sensory organ for hearing perception.

The cochlea has the form of a snail shell with 2.5 turns (Figure 1.3). Inside, there are three fluid-filled chambers, i.e. scala tympani, scala vestibuli (both contain perilymph) and scala media (which contains endolymph). The scala tympani and the scala vestibuli are contiguous, merging at the tip of the cochlea (helicotrema). The scala vestibuli ends at the oval window where the perilymph is stimulated by the stapes; the scala tympani ends at the round window which ensures pressure equalization. In the organ of Corti, which is positioned on the basilar membrane in the scala media, acoustic vibrations are transformed at the level of sensory hair cells by way of a complex mechanism into electrical signals of the cochlear nerve (Figure *1.3*).

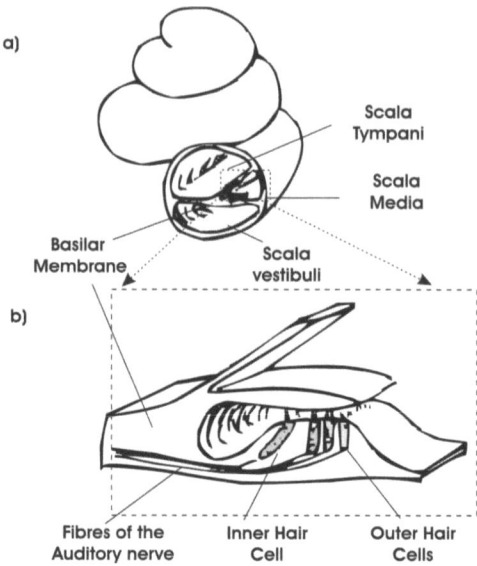

Figure 1.3 *Schematic drawing of the cochlea (a) and the organ of Corti (b) with details of the basilar membrane and hair cells (courtesy M. Kompis)*

The vibrations of the stapes induces a so-called travelling wave ([1] p. 29) on the basilar membrane which shows maximal amplitudes depending on the frequency at different places. Inner hair cells IHC provide the main neural output of the cochlea. They transform mechanical movements into electrical signals. This excitation of the inner hair cells is additionally amplified by outer hair cells (OHC). Outer hair cells act as a non-linear amplifier up to 50 dB for low input levels.

The function of cochlea may be described additionally as a frequency analyser.

1.3 Hearing disorders

Three preconditions must be fulfilled for the perception of acoustical signals. First the sound has to be transmitted from the outer ear to the inner ear fluid, then the sound must be transformed into an electrical signal by the inner (IHC) and outer hair cells (OHC) to the cochlear nerve. The cochlear nerve transmits the electrical signals to the central auditory system of the brain. Diseases can manifest in each of the mentioned parts. For clinical and audiological diagnostic and therapy it is important to associate the hearing problem with their origin. Two kinds of peripheral hearing disorders are principally differentiated, i.e. conductive and sensorineural hearing loss [1], [3].

1.3.1 Conductive hearing loss

In conductive hearing loss, sound is not transmitted efficiently to the inner ear. The origin of conductive hearing loss is located either in the external ear or in the middle ear. The most important causes of persistent conductive hearing loss are [1]:

- Otosclerosis, i.e. pathological ossification with fixation of the stapes in the oval window. [4].
- Chronic otitis media
- Interruption of the ossicular chain
- Malformation dislocation fracture or absence of the ossicles (middle ear)
- Malformation of the external auditory canal with total obstruction of the external ear canal
- Perforation of the tympanic membrane due to an accident such as due to a Q-Tip or to chronic otitis media
- Fracture of the temporal bone resulting from an accident

1.3.2 Sensorineural hearing loss

In sensorineural hearing loss, either transformation of acoustical (mechanical) into neural (electrical) signals in the cochlea is defective, or the transmission via the auditory nerve is not effective or the auditory center is defective. Hearing diseases down stream of the cochlear are defined as retrocochlear hearing loss [3]. Retrocochlear hearing losses are rare and often related to tumors or apoplexy. Therapy with a hearing aid in central auditory dysfunction is principally not possible and therefore not discussed further.

The most important causes of sensorineureal hearing loss are [1]:

- Presbyakusis (hearing loss due to age)

- Congenital (genetically, infection of mother, e.g. rubella)
- Connatal asphyxia (deprivation of oxygen)
- Morbus Menière, Meningitis (disorder of the inner ear)
- Ototoxic hearing loss
- Acoustic trauma:
  - Acute: very high sound pressure levels for a short term
  - Chronic: sound pressure level over a long time (guideline of SUVA: max. 40 hours at 87 dB(A) per week)
- Sudden hearing loss

1.3.3 Combined hearing loss

Combined or mixed hearing loss is a combination of conductive and sensorineural hearing loss.

1.4 Therapies

In the treatment of hearing disorders not only the cause but also the severity need to be evaluated. Hearing loss is described as the difference to normal hearing in dB HL. It may be ranked as mild, moderate, severe or profound. Hearing loss is not necessarily the same over the entire frequency range. Therefore, it is quite common for someone to have more than one degree of hearing loss (e.g. mild sloping to severe).

An overview of therapies for different types and degrees of hearing loss is shown in Figure 1.4. The limits of each therapeutical option are not absolute; e.g. cochlear implants were originally applied for completely deaf patients only (> 110 dB). Today, it penetrates into the field of patients with severe to profound hearing loss with residual hearing at low frequencies.

Maximal hearing loss due to purely conductive hearing loss is about 60 dB i.e. the interruption of the ossicular chain or absence of at least one of the ossicles.

Introduction

In this thesis, implantable hearing systems (IHS I-II) are focussed. For the sake of completeness, all therapies are briefly described subsequently.

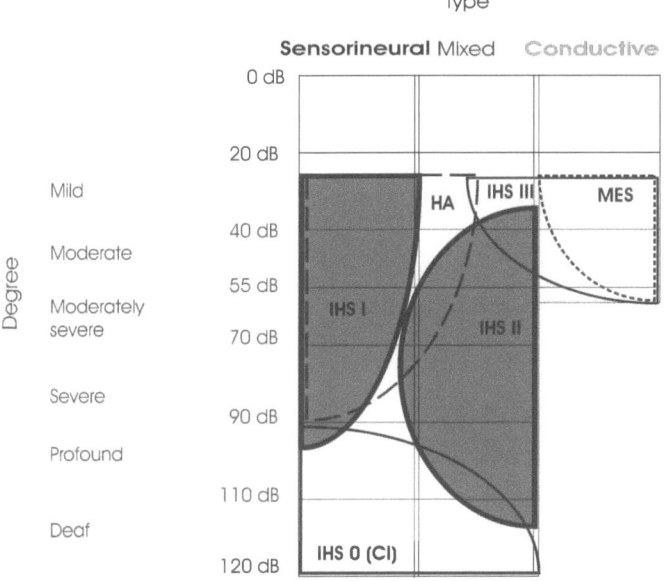

Figure 1.4 *Therapies according to factors type and degree of hearing loss. HA : conventional hearing aids, MES : middle ear surgery, CI: cochlear implants, IHS: implantable hearing systems (Type 0-III).*

1.4.1 Conventional hearing aids (HA)

Conventional hearing aids (HA) represent the state of the art of treatment of sensorineural hearing loss at the present time. They consist of one or more sound collecting microphone(s), a sound processing unit, an amplifier and a loudspeaker. Based on this principle, different types are available, i.e behind the ear (BTE), in the ear (ITE), in the canal (ITC), completely in the canal (CIC).

In the last decade, numerous and significant improvements have been seen in hearing aid technology. Most important are digital hearing aids, multi-microphone systems which offers:

- Different sound processing on multiple channels. Amplification of each channel is calculated according to incoming level and individual hearing loss (multi channel compression).
- Adaptive directive multi-microphone noise reduction which increase speech intelligibility in adverse listening situations and noisy environments.
- Adaptive suppression of the acoustical feedback

Accessories, such as wireless classroom communication systems (FM systems) and remote controls have become increasingly common, facilitating the life of hearing aid users [5].

Conventional hearing aids represent the largest market in hearing impairment with approximately 5.5 million units sold per year [6]. However, approximately 25% - 43% of them are not used [7], [8].

1.4.2 Middle ear surgery (MES)

Conductive hearing loss can be rehabilitated by reconstructing the middle ear or the external auditory canal. (Passive) Middle ear prostheses may be used to replace the defective structures [9].

In the case of otosclerosis (see. 1.3.1), the immobile stapes is partially or totally removed in order to access the inner ear fluid. A stapes prosthesis is then attached to the incus coupled to the fluid of the cochlea at the stage of the oval window (Figure 1.1)

Introduction

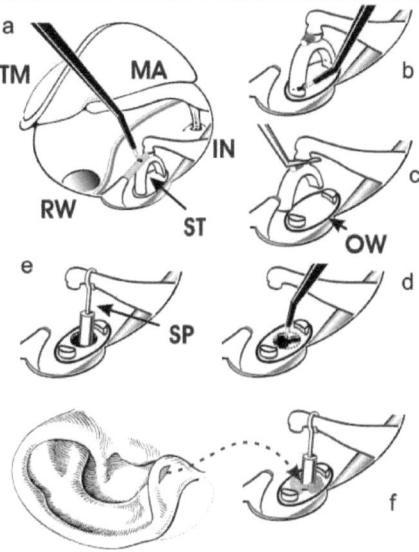

Figure 1.5 Procedure of a stapedotomy. a) access to the tympanic cavity by elevation the tympanic membrane (TM). IN: Incus, Stapes, MA: Malleus, RW: Round window. b,c) Stapes (ST) is partially removed. d) hole in the stapedial footplate provides access to oval window (OW) and the inner ear fluid (perilymph). e) a stapes prosthesis (SP) is crimped on the incus and placed in the oval widow f) the oval widow is sealed with adipose tissue (courtesy R. Häusler)

Stapedectomies are established procedures and performed over 10'000 times per year around the world. A significant improvement of hearing is obtained in over 90 % of operated patients. However, there is a small risk of postoperative deafness (approx.. 1%)[10].

1.4.3 Implantable hearing systems (IHS)

Most of implantable hearing systems, as they exist today, consist of one or more sound collecting microphone(s), a sound processing unit amplifier, and a battery - all worn externally - and an implanted output transducer. The transmission from the external unit to the implant can even be provided by a percutaneous plug or an RF transmission. Depending on the principle of such transducers, they can be applied for different kind of hearing diseases (Figure 1.4).

IHS 0 : Cochlear implants (CI)

Cochlear implants represent one of the most spectacular advances in modern medicine. They render possible that deaf patients with a complete destroyed inner ear but a functional auditory nerve can hear [1],[11]. Congenitally deaf children implanted at an early age achieve not only hearing but also language development [12], [13].

The transducer of a cochlear implant consists of an electrode array, inserted in the cochlea. These electrodes stimulate the auditory nerve electrically.

Cochlear implants replace the function of the cochlea by direct electrical stimulation of the auditory nerve.

IHS I: Implantable middle ear transducers

As mentioned above, conventional hearing aids underwent significant improvements over the past years. Most of them were made with signal processing and related hardware. However, the output signals of the speakers of conventional hearing aids are physically limited to approximately 5000 Hz [14]. For comparison, consonants in spoken language contain frequencies up to 8000 Hz and the range of hearing goes up to 20000 Hz [1].

An important drawback of conventional hearing aids is the occlusion of the external auditory canal. The ear mould must obturate perfectly in order to minimize direct acoustical feedback. However, tight moulds reduce wearing comfort and often lead to termination of use. Additionally, the ear is aerated less well because of the ear mould, which increases the risk for infections of the external auditory canal.

Implantable middle ear transducers could reduce these drawbacks substantially. Therefore, over the last decades many investigations have been made into the development of implantable middle ear transducers [15],[16],[17]. Currently, middle ear transducers are an alternative for patients with problems with conventional hearing aids. To achieve a broad acceptance in the group of conventional hearing aid users (Figure 1.1), they presumably must be implantable minimally invasively and significantly better than conventional hearing aid speakers.

This is not the case for any of existing middle ear transducers (Chapter 2:). Therefore, we investigate in principle a minimally invasive contactless electromagnetic transducer for following reasons:

- Minimally invasive surgery without mastoidectomy decreases the likelyhood of an implantation. (mastoidectomy: opening of the middle ear cavity by drilling a hole in the temporal bone behind the ear)

- No preload on the ossicular chain, which inhibits compensation (Chapter 4)

IHS II: Direct Acoustical Cochlear Stimulation

Currently, there is no effective therapy for severe combined hearing loss. Thus, a transducer for its rehabilitation is innovative and has the potential of occupying a niche. Depending on the design, such a system could even be used for sensorineural or conductive hearing loss alone.

One solution for such a transducer is provided by DACS, an abbreviation for Direct Acoustical Cochlear Stimulation. It works on the principle of direct acoustical stimulation of the inner ear fluid (perilymph). The perilymph is excited by a stapes prosthesis (1.4.2) which is additionally amplified by an actuator (Chapter 5).

Recently, a second principle for direct acoustical cochlear stimulation was presented whereby an IHS I was directly mounted on the round window [18].

IHS III: Bone Anchored Hearing Aids (BAHA)

Bone anchored hearing aids present an alternative therapy to middle ear surgery for conductive hearing loss. In contrast, middle ear surgery, BAHA, can be also used for patients with mild mixed hearing loss [19].

The BAHA consists of a titanium bone screw which is fixed behind the ear and penetrates the skin (percutaneous plug). On this screw, the active part of the BAHA is hooked up. The mechanical output vibrations are directly transmitted to the bone of the scull and then conducted to the cochlea.

1.5 References

[1] Kompis M. Audiologie. Bern: Hans Huber, 2004.

[2] Lang J. Klinische Anatomie des Ohres. Wien: Springer-Verlag, 1992.

[3] Probst R, Grevers G, Iro H, Rosanowski F. Hals-Nasen-Ohren-Heilkunde ein sicherer Einstieg kleine Etappen in Text, Bild und Ton. Stuttgart: Georg Thieme, 2000.

[4] Hausler R, Schar PJ, Pratisto H, Weber HP, Frenz M. Advantages and dangers of erbium laser application in stapedotomy. Acta Otolaryngol 1999;119 (2):207-13.

[5] Kompis M. [New developments in hearing aid technology]. Ther Umsch 2004;61 (1):35-9.

[6] University-Sidney. Hearing aids sold per year.
http://www.usyd.edu.au/research/news/2005/apr/28_fresh.shtml.

[7] Lupsakko AT, Kautiainen JH, Sulkava. R. The non-use of hearing aids in people aged 75 years and over in the city of Kuopio in Finland. Eur Arch Otorhinolaryngol 2005;262 (3):165-9.

[8] WDR. Hearing aid non-users.
http://www.wdr.de/tv/service/geld/inhalt/20030109/b_3.phtml.

[9] Hausler R. [Surgical treatment possibilities of middle ear hearing loss]. Ther Umsch 1993;50 (9):653-62.

[10] Haeusler R. Fortschritte in der Stapeschirurgie. Laryngo-Rhino-Otol 2000;79 Supplement 2:95–139.

[11] Rubinstein JT. How cochlear implants encode speech. Curr Opin Otolaryngol Head Neck Surg 2004;12 (5):444-8.

[12] Manrique M, Cervera-Paz FJ, Huarte A, Molina M. Advantages of cochlear implantation in prelingual deaf children before 2 years of age when compared with later implantation. Laryngoscope 2004;114 (8):1462-9.

[13] Vischer M, Kompis M, Seifert E, Hausler R. [The cochlear implant--evolution of hearing and language with an artificial inner ear]. Ther Umsch 2004;61 (1):53-60.

[14] Knowles. Speakers for Hearing aids. http://www.knowlesacoustics.com/knowlesacoustics-pps/specialty_categorylist.do?category_id=17.

[15] Huttenbrink KB. Current status and critical reflections on implantable hearing aids. Am J Otol 1999;20 (4):409-15.

[16] Chen DA, Backous DD, Arriaga MA, Garvin R, Kobylek D, Littman T, Walgren S, Lura D. Phase 1 clinical trial results of the Envoy System: a totally implantable middle ear device for sensorineural hearing loss. Otolaryngol Head Neck Surg 2004;131 (6):904-16.

[17] Leuwer R. Die apparative Versorgung der Schwerhörigkeit: Konventionelle und implantierbare Hörgeräte. Laryngo-Rhino-Otolo 2005;84 Suppl 1:51-61.

[18] Colletti V, Carner M, Sacchetto L, Colletti L, Giarbini N. The round window approach for vibrant soundbridge. Politzer Society Meeting. Seoul, 2005.

[19] Schupbach J, Kompis M, Hausler R. [Bone anchored hearing aids (B.A.H.A.)]. Ther Umsch 2004;61 (1):41-6.

Chapter 2: Computer assisted optimization of an electromagnetic transducer design for implantable hearing aids

This chapter introduces the minimally invasive implantable middle ear transducer (CLT). It consists of a coil and a magnet to be implanted in the middle ear cavity. In order to optimize the dimensions of the coil, a set of four simulations for different geometries are performed and experimentally verified. As a result, the coil can be optimized either to maximize output levels or to be tolerant of radial displacements between coil and magnet.

Reference:
Stieger C, Wackerlin D, Bernhard H, Stahel A, Kompis M, Hausler R, Burger E. Computer assisted optimization of an electromagnetic transducer design for implantable hearing aids. Computers in Biology and Medicine 2004;34 (2):141-52.

2.1 Abstract

A simple, contactless electromagnetic transducer design for implantable hearing aids is investigated. It consists of a coil and a permanent magnet, both of which are intended for implantation in the middle ear. The transducer is modelled and optimized using computer simulations, followed by experimental verification. It is shown that the proposed transducer design can, because its size and geometry, allow implantation through the external auditory canal and provide a sufficiently high acoustic output corresponding to approximately 120 dB sound pressure level. It can be optimized to be tolerant of radial displacements between coil and magnet of up to 1 mm.

2.2 Introduction

Implantable hearing aids are a dynamic area of research. Several different types of both, totally and partially implantable hearing aids, are currently under development [1,2]. When compared to conventional hearing aids, such implantable aids hold the promise of substantial improvements regarding reduced sound distortion and, consequently, better sound quality and speech recognition, reduced feed-back, better cosmetic appearance and less discomfort due to the occlusion of the ear canal [2].

The single most important component of an implantable hearing aid is the output transducer. It is the equivalent of the loudspeaker in conventional hearing aids but provides a direct mechanical interface in the human middle ear, usually at the ossicular chain. Different types of output transducers have been proposed. Electromagnetic [3] or piezoelectric [4] transducers driving the ossicular chain by means of a driving rod have been shown to be able to provide high output levels of up to 135 dB sound pressure level (SPL) [3]. However, the surgical procedure is usually complex, and additional conductive hearing losses due to the additional mechanical load of the driving rod cannot be ruled out. Electromagnetic floating mass transducers [5] provide only limited output at low frequencies and currently cannot be implanted through the external auditory canal, necessitating a mastoidectomy. Simple electromagnetic transducers consisting of a coil and a permanent magnet have been proposed by several authors [6,7]. This simple contactless design promises several advantages over piezoelectric and more sophisticated electromechanical transducers [8] including:

- reduced risk of malfunctions due to wear and time
- apart from the weight of the part of the transducer which is attached to the ossicular chain, an absence of bias-forces which could cause tissue erosion and promote mechanical device failure
- both parts of the transducer can be implanted or, if necessary, exchanged independently
- minimally invasive implantation through the ear canal, similar to the transcanal approach in middle ear surgery, is probably possible.

The force generated by such contactless electromagnetic transducers depends strongly on geometry and on the relative placing of the components. To our knowledge, the effect of changes in geometry and relative placement have not been investigated systematically. For hearing aid applications, this knowledge is essential in order to insure sufficient acoustic output at acceptable power consumption by optimizing the transducer design. This investigation aims to close this gap for a specific contactless electromagnetic transducer designed to be implanted using a minimally invasive transcanal approach.

The paper is organized as follows. Section 2.3 introduces a simple contactless electromagnetic transducer. In section 2.4, the factors limiting the range of realistic design parameters are discussed. Section 2.5 describes the materials and methods used in the computer simulations and in the experimental verification. Results are presented and discussed in sections 2.6 and 2.7, respectively.

2.3 A simple electromagnetic transducer

Several configurations of electromagnetic transducers for implantable middle ear hearing aids are conceivable and have already been described [6,7]. All of them are based on the principle of a controlled variable force between a coil and a permanent magnet.

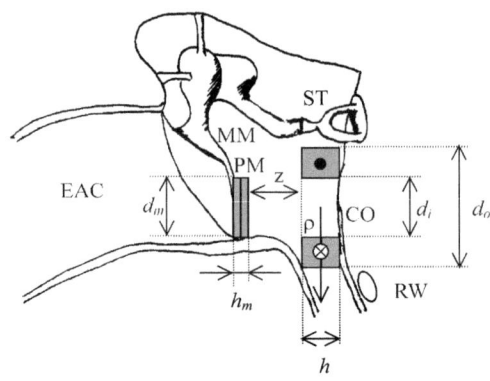

Figure 2.1: Schematic drawing of the proposed electromagnetic transducer for a middle ear hearing aid. A permanent magnet (PM) is mounted on the manubrium mallei (MM). A coil (CO) is placed between the stapes (ST) and the round window (RW) on the wall of the middle ear cavity. (EAC= external auditory canal)

Figure 2.1 shows a schematic representation of the configuration of an electromagnetic transducer considered in this research. The transducer consists of an axially polarized permanent magnet PM mounted at the manubrium mallei MM, and a coil CO mounted between the stapes ST and the round window RW on the part of the surface of the middle ear cavity called promontory. The permanent magnet is defined by its diameter d_m, its height h_m and, in conjunction with the choice of magnetic material, its mass mm and its magnetization. The coil is defined by its outer diameter d_o, its inner diameter d_i, its height h and the number of turns N. Opposing

Optimization

directions of the current flow in the coil are marked by a dot and a crossed circle respectively in Fig. 2.1 and 2.3.

Using these parameters, the cross-section of a single wire q^2 and the total length wire L of the coil can be calculated

$$q^2 = \frac{(d_o - d_i) \cdot \eta \cdot h}{2 \cdot N} \qquad (2.1)$$

$$L = N \cdot \pi \cdot \frac{d_o + d_i}{2} \qquad (2.2)$$

where η is a filling factor between 0 and 1 to account for the space required between wires. Assuming a current I flowing through the coil, the current density j can be calculated as

$$j = \frac{I \cdot N}{\frac{(d_o - d_i)}{2} \cdot h} \qquad (2.3)$$

The current causes a magnetic field H and a magnetic flux density B, which results in a force on the permanent magnet and on the attached ossicle. Mediated by the ossicular chain, this movement is transferred to the inner ear, resulting ultimately in a hearing impression. The force generated by the transducer depends on the relative position and the geometry of the coil and the permanent magnet. This relative position is defined by the radial displacement ρ and the air gap z between the coil and the magnet, as depicted in Fig. 2.1. In our simplified model, the axis of the coil and the magnet are assumed to be parallel at all times.

2.4 Factors limiting the design parameters

A number of factors, including geometry, weight, biocompatibility and tissue warming due to power dissipation of the coil, limit the range of design parameters of electromagnetic transducers.

2.4.1 Geometry

The geometry of the middle ear implies limitation on the size and placement of the coil and the magnet. To illustrate the anatomy and space available in the middle ear cavity, a reconstruction from a series of CT-scans of a temporal bone of a healthy Caucasian adult male is shown in Fig. 2.2. Its size and geometry are typical for adult human middle ears [9, 10]. A coil and magnet, the dimensions of which correspond to the transducer referred to as the reference coil and magnet, are shown in Fig. 2.2. The coil has an outer diameter of 4.3 mm and a length of 0.6 mm, the permanent magnet is 1.5 mm in diameter and 0.5 mm in length. àWengen found that a diameter of 3 mm is reasonable for middle ear implants, but diameters of up to 9 mm might be possible, if a part of the ossicular chain would be removed [11]. A conceptual geometrical study on the range of possible transducer placements and geometries using CT-scans was conducted. Furthermore, coils and magnets of different sizes were actually placed in artificial temporal bones (Pettigrew Plastic Temporal Bones, Stirling, Scotland, UK). It was found that the overall thickness of the complete transducer, including coil, magnet and air-gap must be no larger than 2 mm in order to be implantable in most human middle ears. Assuming a minimal distance between coil and permanent magnet of 0.2 mm, which is considered the lower limit for surgical feasibility, the thickness of the coil and the magnet together must not exceed 1.8 mm. Assuming a cylindrical design, the maximal diameter of the transducer was found to be approximately 7.0 mm.

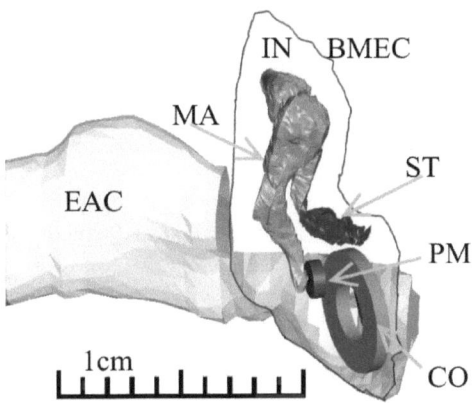

Figure 2.2: 3-D model of external auditory canal (EAC) and the middle ear with the ossicles (malleus MA, incus IN, stapes ST). The transducer, consisting of a coil (CO) and a permanent magnet (PM), lies within the borders of the middle ear cavity (BMEC). The reconstruction is based on a micro-CT scan of a human temporal bone. (Voxel size 0.2 mm for the ossicles, 0.5 mm for the cavum tympani and external auditory canal)

2.4.2 Mass of the permanent magnet

Whenever the implantable hearing aid is turned off, sound transmission is influenced by the mass of the permanent magnet. It has been found that an additional mass of 37.5 mg or 40 mg respectively gives rise to an additional damping of 10 to 20 dB [7,12]. This is unacceptably high for hearing aid applications. For smaller masses between 20 to 25 mg, virtually no additional damping [13] or small additional damping, 1.3 dB [14] and 5 to 10 dB [12], has been reported.

In order to maximize the force generated by the transducer while keeping damping at an acceptably low level, the mass of the permanent magnet, including its casing, should therefore preferably not exceed 25 mg. Biocompatible casing, including the fixation device, has a mass of approximately 10 mg, leaving a maximum of 15 mg for the permanent magnet itself. Therefore, in this study, a SmCo magnet with its diameter 1.5 mm and length of 0.5 mm, and a resulting mass of 14 mg (density $\rho = 8.5$ g/cm^3) was used.

2.4.3 Temperature

Most of the energy transferred to the transducer is transformed into thermal energy at the transducer coil. In order to prevent tissue damage, the surrounding tissue should not be heated by more than 1° C [15]. To determine the amount of energy which can be dissipated by the coil over an extended period of time without excessive warming, a clay model of the temporal bone region, including the middle ear cavity, was build. A coil was placed on one side of the model cavity and heated by means of a constant current of 16 mA (power dissipation 6.4 mW) corresponding to a continuous acoustic output of approximately 120 dB SPL. Although the model did not include blood perfusion, which would normally increase transport of thermal energy away from the coil, the temperature increase of the model tissue directly under the coil did not exceed 0.5° C over an 8 hour period.

2.4.4 Biocompatibility issues

Biocompatibility is a prerequisite for any implantable system. A gas-tight titanium coating of the magnet and the coil could guarantee biocompatibility [16]. With its low relative permeability of $\mu_r = 1.00018$, titanium has essentially no influence on the electromagnetic field.

2.5 Material and methods

2.5.1 Computer simulations

Reference coil

A simulation procedure using Matlab (The MathWorks Inc, Massachusetts, US) was developed to calculate the static force of a coil–magnet configuration with finite dimensions. Calculations were based on the laws of Biot Savart and force calculation on a magnetic dipole in an inhomogenous magnetic field [17]. This procedure has been found to be more efficient than finite element models which are optimized for systems where most of the magnetic flux is guided in metallic materials. In our application, small changes in the choice of meshing have been found to yield substantial differences in the results. Coils were modelled by a number of ideal wires, i.e. wires with infinitely small diameters. To account for the effects of finite coil height and for the difference of the inner and outer diameter, these ideal wires were distributed within the actual coil volume. For each coil, the number of turns was chosen in such a way that the errors in the field distribution were below 1% and the computational load remained reasonable. The current I in each of the simulated ideal wires was calculated according to Eq. 2.3 to match the current density j in the corresponding experiment. For each coil, the relative position of the coil and the magnet, i.e. the parameters ρ and z (cf. Fig. 2.1) were varied systematically in the range of up to 4 mm in either direction for ρ and between 0.1 mm and 1 mm for z respectively.

Variations of the design parameters of the electromagnetic transducer:

Simulations were performed using 5 different coils. In each case, the same wire (i.e. same material, same length L and cross sectional area q^2) and the same filling factor η was assumed, which guarantees the same power consumption for direct current. Assuming these parameters are fixed, h, d_o or d_i can be calculated if the other two are known by correspondingly transforming the equation

$$L = \frac{\pi}{4 \cdot q_{ref}^2}(d_a^2 - d_i^2) \cdot h \cdot \eta \qquad (2.4)$$

A summary of the parameters of the five different coils used is given in Fig. 3. Starting from a reference parameter set defining a reference coil, either the outer diameter or the height of the coil were varied between its minimal and maximal admissible value, adjusting the inner diameter of the coil according to Eq. 2.4. The values for maximal height and maximal diameter were chosen according to the geometric considerations discussed in section 2.4. Minimal values were found by setting the inner diameter of the coil d_i to zero.

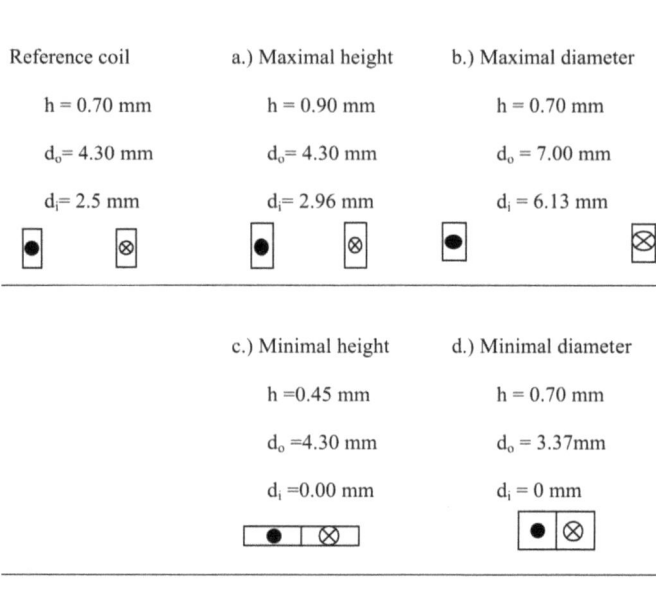

Figure 2.3: Geometries of the coils used in the simulations. (Dimensions of schematic drawings for illustration purposes only and not to scale.)

2.5.2 Experimental verification

To verify the simulation results, static force measurements of the transducer with the reference coil were performed. Isolated copper wire with a diameter of 40 µm was used. The total length of the wire of the coil was 1.75 m (cf. Eq. 2.2), resulting in 166 turns of the coil and a filling factor η of 0.35.

The force generated by the transducer was measured using a Mettler Toledo Type AG 204 scale. The resolution of this scale is 0.1 mg, corresponding to a force resolution of approximately 0.92 µN. To avoid interactions between the transducer and the scale, the coil was separated from the surface of the scale using a 2 cm block of synthetic material (PVC) to which the coil was attached. A constant current of 16 mA was applied using a Keithley Source Meter 2400, which resulted in a current density j of 4.46 A/mm^2.

The magnet was mounted on an x-y-z displacement system (Newport 4-Axis Motion Controller MM 4005). The magnet was moved into a field of 5,6 mm x 5,6 mm x 2,4 mm using a step size of 0.2 mm. The measurement was automated using a LabView program which was used to drive the displacement system and for data acquisition from the scale. After each displacement of the magnet, the system was allowed to stabilize for 6 s, resulting in an approximate total measuring time of 18 hours.

2.6 Results

2.6.1 Computer simulations

Reference coil

The left-hand diagram in Fig. 2.4 shows the axial force as a function of the width of the air gap z and the radial displacement ρ of the permanent magnet. The right-hand side of Fig. 2.4 shows the vector force field, i.e. the axial and the radial component, as a function of the radial displacement of the magnet at a fixed air gap of 0.2 mm.

For any constant radial displacement r, force decreases for larger air gaps z. For small air gaps (z < 0.6 mm), there are two separate maximums as a function of the radial displacement ρ. For larger air gaps, the two maximums merge into a single, initially broad maximum. The maximal force F_{tot} = 0.52 mN is obtained at ρ = 0.8 mm and the shortest air gap considered, z = 0.1 mm.

The right hand side of Fig. 2.4 shows that the force is purely axial in the center of the coil (ρ= 0 mm). For any other ρ, there is a radial component, which is greatest above the edge of the coil (ρ= 1.8 mm).

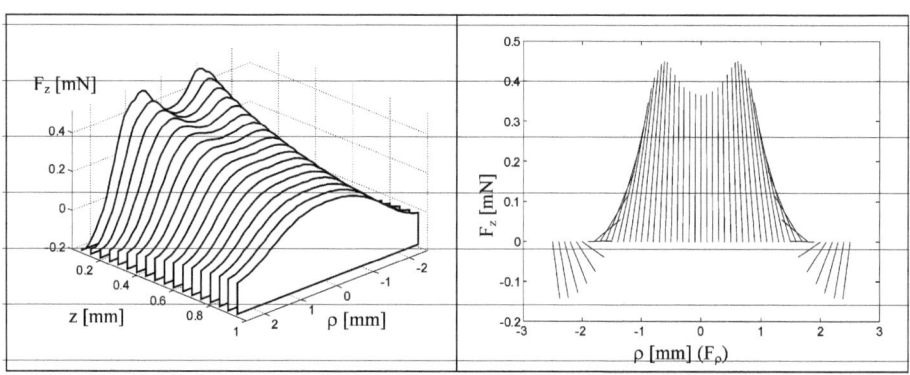

Figure 2..4: Left: Force Fz in z direction in dependence of the air gap z and the radial displacement ρ of the reference coil. Right: Vector of the total force F_{tot} for a constant air gap z = 0.2 mm of the reference coil. The component scale for the length of the force F_ρ is the same as for the perpendicular force F_z on the ordinate. The vectors start on the abscissa (Fz = 0) at the point of their lateral displacement ρ.

Variations of the design parameters of the electromagnetic transducer

Fig. 2.5 shows the results of the simulations of the axial force for the four other coils with maximal and minimal coil diameters and coil height, respectively. For coils a.) and b.) (maximal height and maximal diameter), the force as a function of the radial displacement ρ and the air gap shows qualitatively a similar behavior as the reference coil: there are two separate maximums for small air gaps z which merge into a single, relatively broad maximum at larger z. However, compared to the reference coil, the maximal force is 12% smaller in case a.) (maximal height) and by a factor of 2.0 in case b.) (maximal diameter).

For coils c.) and d.) (minimal height and minimal diameter) there is only a single maximum at any air gap z for the given set of parameters. For short air gaps, i.e.

Optimization

$z = 0.1$ mm, the maximal force is larger by a factor of 1.9 and 2.2 respectively than for the reference coil.

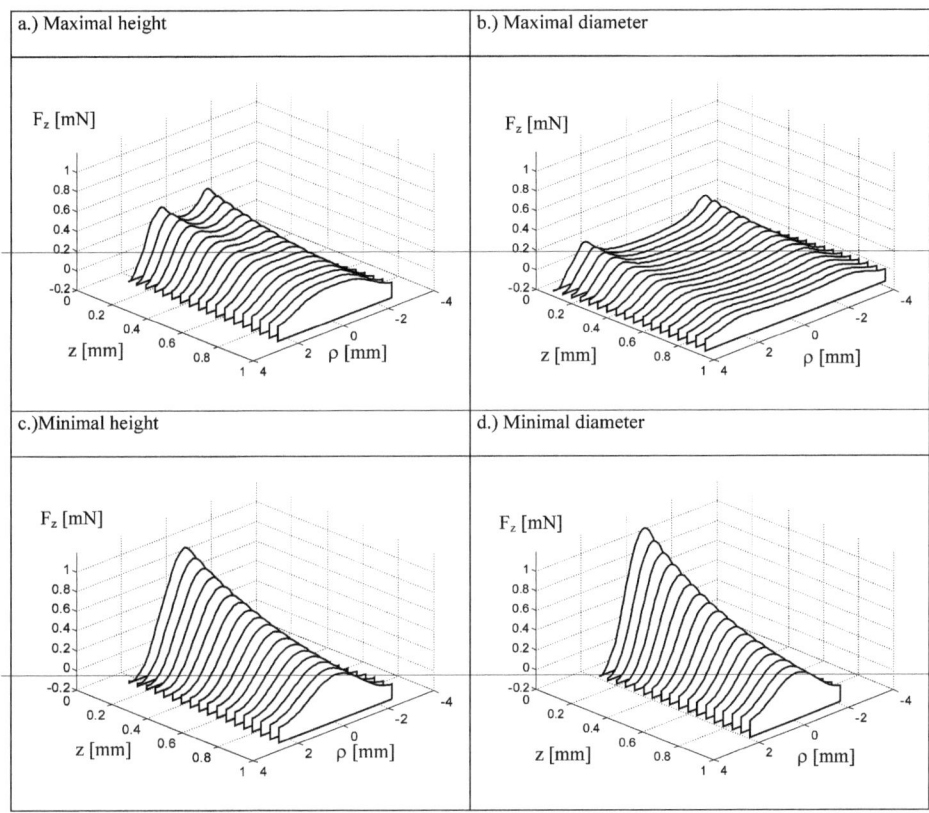

Figure 2.5: Results of simulations for 4 different coil geometries, as listed in Fig. 2.3.

2.6.2 Experimental verification

Figure 2.6 shows a comparison of the axial forces for air gaps $z = 0.2$ mm through 0.8 mm in steps of 0.2 mm for the reference coil. The results of the simulations (c.f. Fig. 2.4) and the measurements are plotted in the same diagrams to facilitate comparison. A good qualitative and quantitative agreement, for the most part within a few percents, can be observed. The largest difference of 15% is observed at $z = 0.2$ mm and $r = 0.8$ mm. There is a small asymmetry in the experimental results with respect to radial displacement, which cannot be found in the simulations and is probably due to a small deviation of the relative axes of the coil-magnet system from the axes of the x-y-z displacement system during measurements.

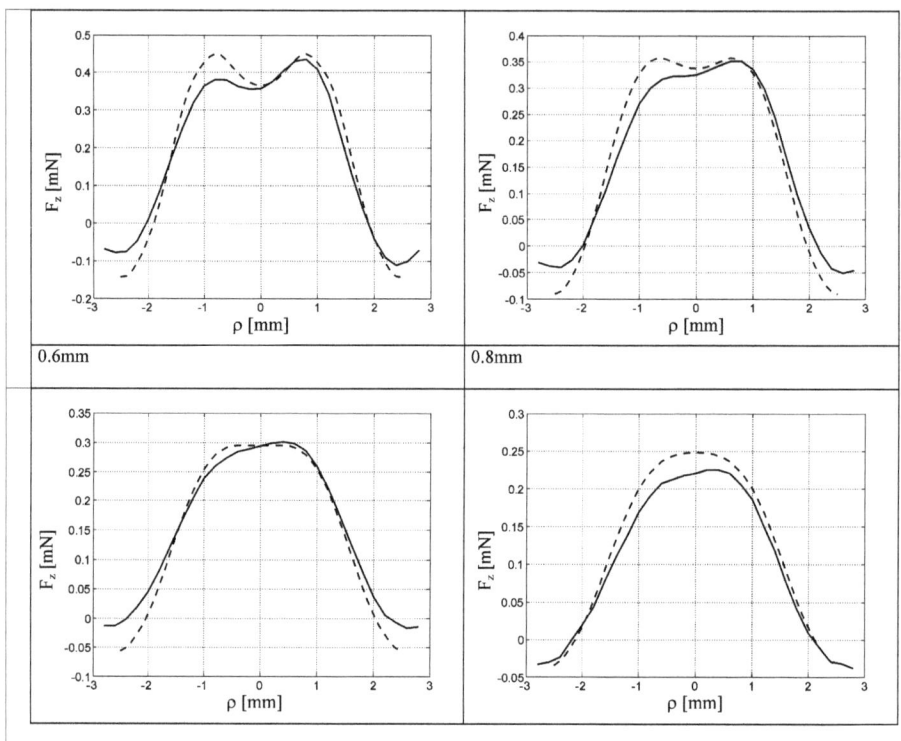

Figure 2.6: Comparison of the results of the computer simulations (dashed lines) and the actual experiments (solid lines) for air gaps of 0.2 mm, 0.4 mm, 0.6 mm and 0.8 mm.

2.7 Discussion

For the reference coil, the force generated by the transducer was both simulated and verified experimentally. The results of our computer simulations and of the experiments are in good agreement, confirming the correctness of the computer simulation approach used. The actual time to perform a complete set of simulations (i.e. the data for one coil, as shown in Fig. 2.4) is approximately 70 hours on a Sun Ultrasparc 10 Workstation. This is substantially longer than the actual measuring time of 18 hours for a measurement. However, sixteen times more data have been covered per simulation than in the measurement, and preparing a new experiment using a different coil is considerably more time consuming than preparing a new set of simulations. Therefore, we prefer the simulation method for systematic parameter variations.

A practically useful implantable transducer should be able to generate forces corresponding to acoustic sound pressure levels of at least 110 to 120 dB SPL. According to data presented by Maniglia [6], 120 dB SPL corresponds to forces of around 0.9 mN. For the transducers considered in this research, this value can be reached only by two out of five sets of design parameters, stressing the importance of computer aided optimization.

In principle, the generated force by the transducer can be increased by increasing the current through the coil. For currents in the order of magnitude of 16 mA (P=6.4mW) considered in this research, tissue warming is not an issue. However, total power consumption and, consequently, the battery life of the hearing aid, may be severely compromised by higher currents. For piezoelectric transducers, power requirements of 2.5 mW for an output corresponding to 120 dB SPL at 1000 Hz was reported [18], which is reasonable for implantable hearing aids. However, if an increase in power consumption by a factor of 10 should become necessary, it would compromise the usefulness of the electromagnetic transducer significantly.

The highest output is reached by the two coils featuring minimal coil height and minimal coil diameter. These are the only two coils which are completely filled with wire, i.e. where there is no air left in the center of the coil. As a consequence, the maximal force is higher by a factor of 1.9 or 2.2 respectively than for the reference coil. As a drawback, the setting is more sensitive to lateral displacements between coil and magnet. A radial displacement of $\rho = 1$ mm causes a drop by 6.2 dB (coil with minimal height) or 8.8 dB (coil with minimal diameter). In contrast, for the reference coil, at an air gap z of 0.5 mm, the drop for a radial displacement of 1 mm is virtually zero. In other respects, all coils behave similarly. Most important, an increase in the air gap z from 0.2 to 0.5 mm results in a comparable drop in force by 3 to 4.4 dB.

Therefore, two different strategies can be pursued in order to optimize the transducer. Maximal force can be optioned by filling the coil with wire all the way to its center. If an air space is left open in the center, maximal force drops by a factor of approximately 2, however, the system becomes tolerant to radial displacements by up to approximately 1 mm.

2.8 Summary

A contactless electromagnetic transducer design for implantable hearing aids, consisting of a coil and a permanent magnet, is investigated. The transducer is intended to be implanted in the middle ear using a minimally invasive surgical procedure through the external auditory canal. The transducer is investigated and optimized using computer simulations for different coil designs and for a range of radial displacements and air gaps. A subset of the simulation results are verified experimentally in a laboratory setting. Results from the simulations and the experiments were found to be in reasonable agreement.

It is shown that the proposed transducer design can, at a size and geometry which should allow an implantation through the external auditory canal, provide an acoustic output corresponding to 120 dB SPL. It is shown that the transducer can be optimized either to maximize output levels or to be tolerant of radial displacements of up to 1 mm between coil and magnet.

2.9 References

[1] K.B. Hüttenbrink, Current status and critical reflections on implantable hearing aids, Am J Otol 20(4) (1999) 409-15.

[2] A.J. Maniglia, D.W. Proops (Eds.), Implantable electronic otologic devices: State of the art, Otolaryngol Clin North Am 34(2), 2001.

[3] J.F. Kasic, J.M. Fredrickson, The otologics MET ossicular stimulator, Otolaryngol Clin North Am 34(2) (2001) 501-514.

[4] H.P. Zenner, H. Leysieffer, Total implantation of the Implex TICA hearing amplifier implant for high frequency sensorineural hearing loss: the Tubingen University experience, Otolaryngol Clin North Am 34(2) (2001) 417-446.

[5] R. Junker, M. Gross, I. Todt, A. Ernst, Functional gain of already implanted hearing devices in patients with sensorineural hearin loss of varied origin end extent: Berlin experience, Otol Neurolotol 23 (2002) 452-456.

[6] W.H. Ko, W.L. Zhu, A.J. Maniglia, Engineering principles of mechanical stimulation of the middle ear, Otolaryngol Clin North Am, Vol. 28(1) (1995) 29-41.

[7] J.D.V. Hough, R.K. Dyer, K.J. Dormer, P. Matthews, R.Z. Gan, M.W. Wood, Middle ear electromagnetic implantable hearing device, in: J.J. Rosowski, S.N. Merchant (Eds.), The function and mechanics of normal, diseased and reconstructed middle ears, Kugler Publications, The Netherlands, 2000, pp. 353-366.

[8] M. Kompis, C. Kuhn, M. Affolter, U. Brugger, R. Häusler, Design considerations for a contactless electromagnetic transducer for implantable hearing aids, Proc Annu Int Conf IEEE Eng Biol Soc, 20(6), 1998, pp. 3173-3176.

[9] B.J. Anson, Surgical anatomy of the temporal bone, Raven, New York, 1992.

[10] A.J. Gulya, H.F. Schuknecht, Anatomy of the temporal bone with surgical implications, Parthenon, York, 1995.

[11] D.F. àWengen, Basis for a new semi-implantable electro-magnetic middle ear hearing aid, Habilitation, Basel, Switzerland, 1998.

[12] S. Nishihara, R.L. Goode, Experimental study of the acoustic properties of incus replacement prostheses in a human temporal bone model, Am J Otol 15(4) (1994) 485-94.

[13] F.M. Snik, W.R. Cremers, The effect of the "floating mass transducer" in the middle ear on hearing sensitivity, Am J Otol 21(1) (2000) 42-48.

[14] G.Z. Gan, R.K. Dyer, M.W. Wood, K.J. Dormer, Mass loading on the ossicles and middle ear function, Ann Otol Rhinol Laryngol, 110 (2001) 478-485.

[15] G.M.J. Van Leeuwen, J.J.W. Lagendiik, B.J.A.M. Van Leersum, A.P.M. Zwanborn, S.N. Hornsleth, A.N.T.J. Kotte, Calculation of change in brain temperature due to exposure to a mobile phone, Phys Med Biol, 44 (1999) 2367-2379.

[16] E. Wintermantel, H. Suk-Woo, Biokompatible Werkstoffe und Bauweisen, Springer, Berlin, 1998.

[17] F.K. Kneubühl, Repetitorium der Physik, Teubner, Physik, Stuttgart, 1990.

[18] H. Leysieffer, J.W. Baumann, G. Müller, H.P. Zenner, Ein implantierbarer piezoelektrischer Hörgerätewandler für Innenohrschwerhörige, Teil II: Klinisches Implantat, HNO 45 (1997) 801-815.

Chapter 3: Anatomical study of the human middle ear for the design of implantable hearing aids

In this chapter a set of morphometric data is generated which is mandatory for the design of the CLT, DACS and also for others IHS which requires the space available in the middle ear cavity. The data was generated using computer tomography (CT) scans of human heads postmortem. These data were statistically examined on different factors.

Reference:

Stieger C, Djeric D, Kompis M, Remonda L, Hausler R. Anatomical study of the human middle ear for the design of implantable hearing aids. Auris Nasus Larynx 2006;33 (4): 375-80.

3.1 Abstract

Objective: To generate anatomical data on the human middle ear and adjacent structures to serve as a base for the development and optimization of new implantable hearing aid transducers. Implantable middle ear hearing aid transducers, i.e. the equivalent to the loudspeaker in conventional hearing aids, should ideally fit into the majority of adult middle ears and should utilize the limited space optimally to achieve sufficiently high maximal output levels. For several designs, more anatomical data are needed.

Methods: Twenty temporal bones of ten formalin-fixed adult human heads were scanned by a computed tomography-system (CT) using a slide thickness 0.63mm. Twelve landmarks were defined and 24 different distances were calculated for each temporal bone.

Results: A statistical description of 24 distances in the adult human middle ear which may limit or influence the design of middle ear transducers is presented. Significant inter-individual differences but no significant differences for gender, side, age or degree of pneumatization of the mastoid were found. Distances, which were not analyzed for the first time in this study, were found to be in good agreement with the results of earlier studies.

Conclusion: A data set describing the adult human middle ear anatomy quantitatively from the point of view of designers of new implantable hearing aid transducers has been generated. In principle, the method employed in this study, using standard CT-scans, could also be used preoperatively to rule out exclusion criteria.

3.2 Introduction

Extensive work on the anatomy of the human ear has been published [1, 2, 3]. Furthermore, the dimensions of the ossicles or parts of these, such as the superstructure of the stapes, have been studied in detail [4, 5].

Nevertheless, data on the dimensions of the free space in the middle ear, especially quantitative data on distances which are not important in conventional middle ear surgery, are rare. The knowledge of such data is important when designing middle ear transducers for totally or partially implantable hearing aids.

Middle ear transducers, i.e. the equivalent of the loudspeaker of conventional hearing aids, are the most important part of implantable hearing aids [6]. Middle ear transducers are designed to be placed either partially or completely in the middle ear cavity and should ideally be implantable in any adult patient's ear. However, this is not always the case. Massen et al. [7] reported that for one specific design (TICA [8]) implantation was not possible in 11 out of 50 human temporal bones due to anatomical constraints. They proposed an X-Ray and CT based method where the distance between the sinus sigmoideus and the posterior wall of the external auditory canal was measured in order to evaluate implantability preoperatively. Similarly, Esselmann et al. [9] and Dammann et al. [10] proposed CT based surgical planning and test fitting procedures for two different implantable hearing aids (TICA [8] and MET [6]) using 3D-reconstruction. However, overall time for a single reconstruction amounted to 4-6 hours [9] or 50 min [10] respectively. These studies indicate that a criterion for testing implantability preoperatively for implantable middle ear transducers is in demand. More importantly, they emphasize the importance of the availability of reliable anatomical data during the design phase of any new implantable hearing aid transducer.

The primary focus of this study was to determine a set of human ear lengths and ranges of the most important distances, which influence the design of implantable middle ear transducers. In addition, the method for these measurements was chosen in a way that makes possible direct applicability *in vivo*, so that it can be used to verify implantability before surgery in the future.

3.3 Material and Method

Twenty temporal bones of ten formalin-fixed human heads (4 females and 6 males) with an average age of 77.4 years (range 69 to 88 years) were provided by our university's Department of Anatomy. In this study, the bones were obtained from donors with no known otological problems. The heads were scanned axially with a multidetector row (8) CT (Light Speed Ultra, GE MedicalSystems, Milwaukee, Wis, USA). The following standard parameters for clinical temporal bone examination were applied: kilovoltage setting of 120 kV; tube current, 160 mA; collimation 4 x 1.25 mm; table feed, 5 mm per rotation; rotation time 600 ms. Secondly, the slide thickness and the increment from slice to slice were reduced to 0.625 mm. Each voxel was measured 0.2 mm x 0.2 mm x 0.625 mm. Two volumes of 10 cm x 10 cm x 5 cm centered around the stapes of each ear were analyzed.

In order to reduce the actual complexity of the anatomy of the middle ear, we defined 12 landmarks, numbered 1 through 12, which are relevant for the design of middle ear transducers (Fig. 3.1).

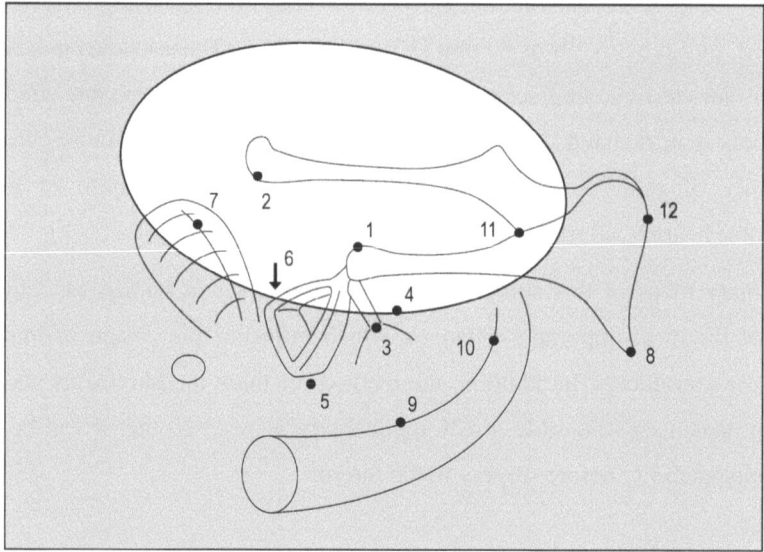

Figure 3.1: Schematic view of the middle ear with facial nerve and definition of the anatomical landmarks

Morphometric data

1. Tip of lenticular process of incus (processus lenticularis incudis)
2. Tip of manubrium mallei
3. Pyramidal eminence
4. Posterior border of annulus
5. Posterior border of oval window niche
6. Anterior border of oval window niche
7. Highest point of promontory (shortest distance to manubrium mallei)
8. Short process of the incus (crus breve incudis)
9. Vertical part of facial canal (shortest distance to posterior border of annulus)
10. Second genu of facial canal
11. Most inferior part of the articulation of malleus and incus
12. Top of the corpus incudis

The coordinates of each landmark were measured using ImageJ Software, version 1.30 (National Institutes of Health, USA, freeware). Nine of these landmarks were directly and unequivocally identifiable in the CT images (Fig. 3.2). Three landmarks, i.e. 4, 7, 9, required each a selection out of multiple points (e.g. 3.4a, 3.4b) in such a way that a given distance (2-7 or 4-9 respectively) was minimized (Fig. 3.2).

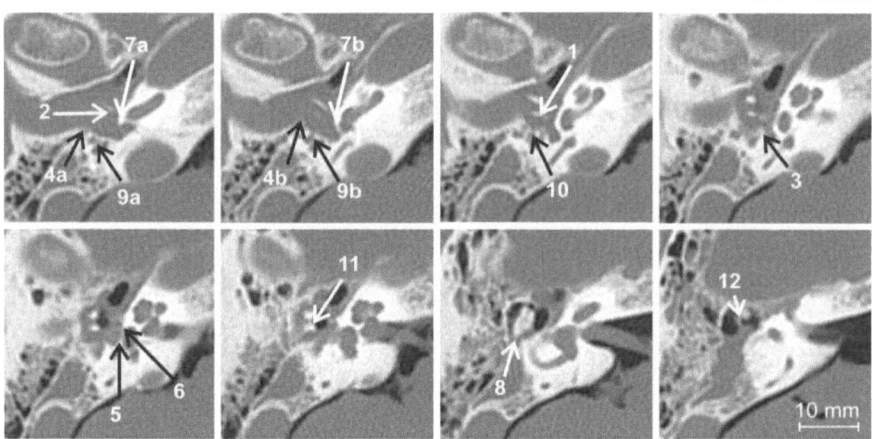

Figure 3.2: Typical subset of slices out of a right ear CT scan used for identification of landmarks 1-12. Slices are arranged in increasing order (inferior to superior) starting in the top left image. Landmarks 4, 7, 9 required a selection of multiple points (i.e. a, b) to find the minimal distance between 2-7 or between 4-9.

Theoretically, 66 distances can be determined between any given 12 landmarks. We confined ourselves to a subset of 24 important distances, marked with capital letters A through X (Table 3.1), which may reasonably be expected to have an impact on the development of implantable middle ear transducers [11]. All distances were calculated automatically using Mathcat Software (Mathsoft Inc, Cambridge, MA, USA). Statistical analysis was performed with R: a language and environment for statistical computing (R Foundation for Statistical Computing, Vienna, Austria) and SPSS (Version 13.0.1, Chicago IL, U.S.A.). As normal distributions could not be assumed, unpaired Wilcoxon rang sum tests were used to find 95% confidence intervals of the median for differences between males and females. Paired Wilcoxon rang sum tests were used to analyze left-to-right differences, respectively. A scale with four degrees of pneumatization (DP1-4) [7] was used to classify each temporal bone. DP1 denotes a mastoid with exclusively little and small cells, DP2 a mastoid with predominantly small cells, DP3 a mastoid with predominantly large cells and DP4 a mastoid with exclusively large cells [7]. Series of Jonckheere Terpstra tests (SPSS, 2-tailed Monte Carlo significance) were performed to quantify the dependency of each anatomical distance considered in this study with the degree of pneumatization. 95% confidence intervals of the linear regression coefficients of the distances versus the age of the subjects were calculated. No Bonferroni correction was used for these statistical analyses, as discussed below.

3.4 Results

All CT scans were reviewed by an experienced neuro-radiologist and an experienced ENT surgeon and were found to be of flawless quality and showed no abnormal anatomies of the temporal bone. Table 3.1 shows mean values, medians and 95% confidence intervals of the medians for all examined distances A-X.

Figure 3.3 shows a graphical representation of the data, giving quartiles, minima and maxima for each of the distances. For most distances, the difference between the 1st and the 3rd quartile is in the order of magnitude of 1 mm, with most absolute lengths lying between 2 mm and 8 mm. However, for several distances (e.g. D, Q, V, X), the extreme values are considerably (more than a factor of 2) farther apart than the distance between the 1st and 3rd quartile.

For several contactless electromagnetic transducer designs employing a moving permanent magnet [e.g. 11, 12], probably the most important distances are B, E, F and G. Distance B, the minimal distance between the tip of the manubrium mallei to the promontory, is considerably smaller than E, the distance between the promontory and the tip of the long process of the incus. This holds true for all quartiles. The smallest measured value was 1.3 mm for B and 2.7 mm for E. Likewise, distance F, i.e. the distance between the tip of manubrium mallei and the tip of long process of the incus is considerably smaller than G, the distance between the tip of the long process of the incus and the posterior border of the middle ear cavity for all quartiles. Minimal measured values were 2.0 mm for F and 3.7 mm for G.

Table 3.1: Definition of 24 distances A through X and statistical distribution.

Distance Label	Endpoints Landmarks	Mean (mm)	Median (mm)	95 % confidence interval of median (mm)
A	1-12	6.3	6.3	6.1-6.6
B	2-7	2.1	1.9	1.7-2.3
C	4-9	3.3	3.3	3.1-3.6
D	1-3	2.1	2.2	1.9-2.3
E	1-7	3.6	3.6	3.3-3.8
F	1-2	3.1	3.0	2.7-3.4
G	1-4	5.1	5.0	4.7-5.5
H	2-4	6.2	6.1	5.9-6.5
I	1-5	4.1	4.2	3.9-4.3
J	1-6	4.0	3.9	3.7-4.1
K	1-11	3.2	3.2	3.0-3.4
L	2-5	4.8	4.7	4.4-5.2
M	2-11	5.0	5.1	4.8-5.2
N	3-5	4.2	4.1	3.9-4.4
O	3-6	3.0	2.8	2.8-3.2
P	3-7	4.6	4.5	4.3-4.9
Q	3-10	3.3	3.2	3.0-3.7
R	4-6	7.9	7.8	7.6-8.3
S	4-7	7.1	7.1	6.7-7.5
T	5-7	3.6	3.7	3.3-3.9
U	6-7	4.4	4.2	4.0-4.7
V	6-9	7.7	8.0	7.1-8.2
W	8-10	3.8	3.7	3.6-4.0
X	9-10	4.1	4.1	3.7-5.1

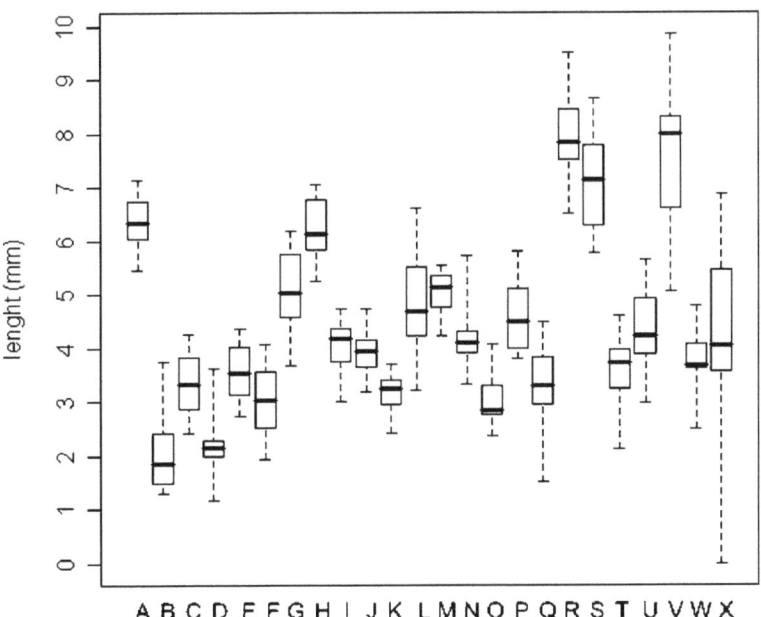

Figure 3.3: Distribution of all 24 distances A through X. Boxplots denote quartiles as well as minimal and maximal values.

Figure 3.4 top shows the differences between males and females for each of the 24 distances; figure 4 bottom shows a similar representation for left-to-right differences. In both figures, a value of 0 mm denotes identical lengths across gender or side, respectively. The tick inside each bar indicates the difference of the median distances; the length of the bar corresponds to the 95% confidence interval of the differences.

In figure 3.4 top, all 95% confidence intervals include the value 0 mm, showing no significant differences between genders. All medians are in the range of -0.4 mm to

0.5 mm and the length of the confidence intervals are in the range of 0.3 mm - 1.7 mm.

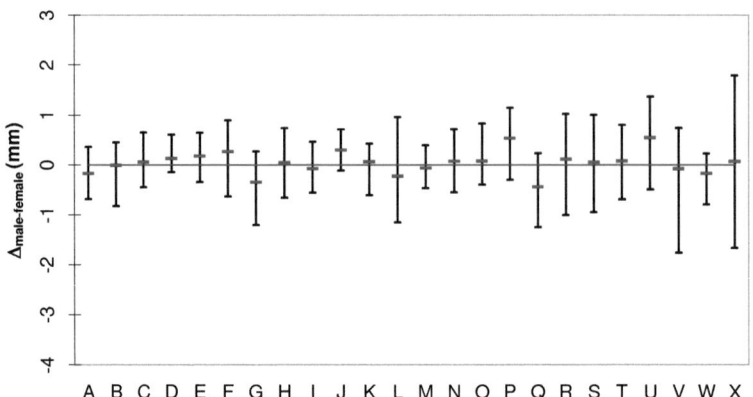

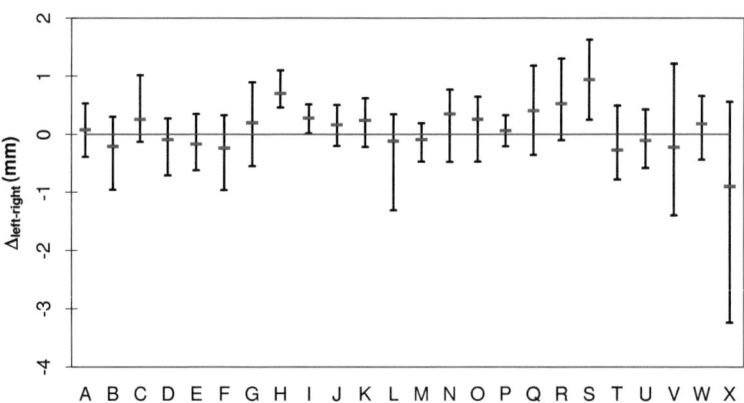

Figure 3.4: Median values and 95% confidence intervals. Top: Differences between male and female subjects. Bottom: Individual left-to-right differences.

In figure 3.4 bottom, two confidence intervals, H and S, exclude 0 mm. The differences are 0.7 mm for H and 0.9 mm for S. The lengths of the confidence intervals are in the range of 0.2 mm - 2.3 mm.

The degree of pneumatization (DP) as described by Maassen et al. [7] was determined for all temporal bones. One temporal bone was classified as DP1, seven temporal bones as DP2, nine temporal bones as DP3 and three temporal bones as DP4. Two-tailed Jonckheere-Terpstra test showed a significant ($p < 0.05$) trend only for the distance I and no significant trends for the other 23 distances. The single significant trend found was negative, i.e. distances I tend to decrease with increasing DPs (Fig. 3.5).

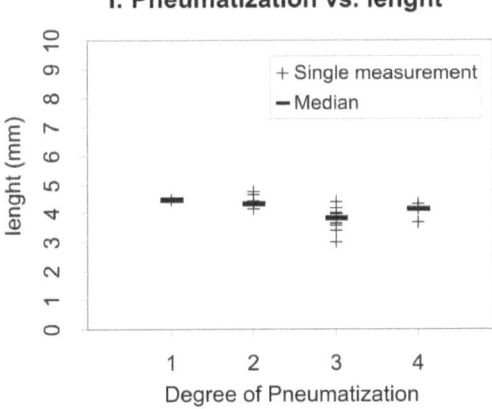

Figure 3.5: Distribution of the values for Distance I versus degree of pneumatization.

Regression correlation coefficients for each of the 24 distances A through X versus age was calculated and yielded values between -0.06 to $+0.08$. The 95% confidence interval for the largest correlation coefficient was -0.02 to 0.15, suggesting no relevant age dependency for the range of ages included in this investigation.

3.5 Discussion

The main purpose of this study was to describe the anatomy of the human middle ear quantitatively in such a way that it can serve as a sound basis for the development of novel middle ear transducers for implantable hearing aids.

The most restraining distances for several types of contactless electromagnetic transducer designs employing a moving permanent magnet [e.g. 11, 12] are the minimal distance between the tip of the manubrium mallei and the promontory B, between the promontory and the tip of the long process of the incus E, between the tip of the manubrium mallei and the tip of long process of the incus F, and between the tip of the long process of the incus and the posterior border of the middle ear cavity G. These distances directly limit the maximal volume of such transducers and therefore influence the maximal acoustic output.

As B was smaller than E through all quartiles, B, that is, the distance between the tip of the manubrium mallei and the promontory, limits the maximal height of this type of transducer. Our data suggests that this dimension should not exceed 1.3 mm if the transducer uses the entire space between the promontory and the manubrium mallei.

On the other hand F, i.e. the distance between the tip of the manubrium mallei and the tip of long process of the incus, is considerably smaller than G, the distance between the tip of the long process of the incus and the posterior border of the middle ear cavity. Therefore, F rather than G is the limiting distance for the design of a permanent magnet attached to the incus [11]. The minimum was found to be just below 2 mm, which led to a magnet radius of 1.7 mm for the design described in Stieger et al. [11].

However, these relations show only the application of the measured data to one specific type of transducer. Most of the other 22 distances are potentially useful for the design of other concepts of implantable hearing aids. Furthermore, our statistical analysis showed that the distances in middle ear are basically independent of gender, side, degree of pneumatization or age within the range considered. Nevertheless, our data suggest that there is a statistically significant left-to-right difference for the distances H and S (Fig. 3.4). However, when calculating 48 different 95% confidence intervals (cf. Fig. 3.4), statistically, approximately 2 of them (5% of 48) can be expected to fall out of the interval, even if there is no statistically significant difference. Additionally, our data suggest decreasing distance I for increasing degree of pneumatization. Again, such a result can be expected if tests with 24 variables are performed and no Bonferroni correction is made (Fig. 3.5).

Because formalin fixed human temporal bones have been used for this study, CT images with a relatively high resolution could be obtained without any exposure to living human subjects. However, formalin shrinks some soft tissue considerably, e.g. by approximately 8% for the human brain compared to living tissue [13]. We could not find data on the influence of formalin on the distances of the middle ear. However, a limited comparison of our results with the results of in vivo measurements by Luntz et al [14] is possible. The distance between the tip of the short process of the incus and the tympanic segment of the facial canal W covers the range of 2.5 to 4.8 mm (average 3.8 mm) in our study and is very similar to their data (range 2.0-4.8 mm, average 3.3 mm).

Our results found by CT are reasonably consistent with those found in quantitative anatomical studies using extraction of individual ossicles. The total height A of the incus in our study (range 5.2-7.1 mm, average 6.3 mm) is similar to the findings by Olszewsky et al. [5] (average 7.21 ± 0.19 mm standard deviation) and that found by Kikuchi [15] (range 5.4-7.0 mm, average 6.5 mm).

Sorensoen et al. [16] presented a method to generate temporal bone data with a much better resolution (voxel size 50 µm). However, such a high accuracy does not seem to be justified for the development of middle ear transducers and could not be reproduced in vivo to assess implantability preoperatively. As the ossicles in the middle ear move due to the static pressure in the order of 1 mm [17], the accuracy of our data of 0.68 mm, which corresponds to the length of the diagonal of one voxel, seems sufficient and reasonable. It should be noted that this accuracy has been obtained using standard CT-equipment and a standard CT-protocol, which could be directly adapted to a simple pre-operative evaluation of patients. In contrast to computer-aided surgical planning tools, where manual segmentation is still necessary [9, 10], this method requires only the determination of a small number of points in the CT scans.

3.6 Conclusions

A data set describing the adult human middle ear anatomy quantitatively from the point of view of designers of new implantable hearing aid transducers has been generated. In principle, the method employed in this study using standard CT-scans could also be used preoperatively to rule out exclusion criteria.

3.7 References

[1] Djeric D, Savic D (1987) Anatomical characteristics of the fossula fenestrae vestibuli. J Laryngol Otol 101 : 426-431.

[2] Donaldson JA, Duckert LG, Lambert PM, Rubel EW (1992) Surgical anatomy of the temporal bone. 4th edn. Raven Press, New York.

[3] Lang J (1992) Klinische Anatomie des Ohres. [in German] Springer, Wien New York.

[4] aWengen D, Nishihara S, Kurkokawa H, Godde RL (1995) Measurements of the stapes superstructure. Ann Otol Rhinol Laryngol 104 : 311-316.

[5] Olszewsky J, Latkowsky B, Zalewski P (1987) Morphométrie des osslets de l'ouie pendent le dévelopement individuel de l'homme et son utilié pour cophochirurgie [in French]. Les Cahiers d'ORL 22 : 488-492.

[6] Hüttenbrink KB (1999) Current status and critical reflections on implantable hearing aids. Am J Otol 20(4) : 409-415.

[7] Maassen MM, Lehner R, Dammann F, Lüdtke R, Zenner HP (1997) Der Stellenwert der konventionellen Röntgendiagnostik nach Schüller sowie der Computertomographie des Felsenbeins bei der präoperativen Diagnostik des

Tübinger implantierbaren Kochleaverstärkers. [in German] HNO 46(3) : 220-227.

[8] Zenner HP, Maassen MM, Lehner RL, Baumann JW, Leysieffer H (1997) Ein implantierbares Hörgerät für Innenohrschwerhörigkeiten – Kurzzeitimplantation von Mikrophon und Wandler". HNO 45 : 872-880.

[9] Esselmann GH, Coticchia JM, Wippold FJ, Fredrickson JM, Vannier MW, Neely JG (1994) Computer-simulated test fitting of an implantable hearing aid using three-dimensional CT scans of the temporal bone: preliminary study. Am J Otol 6 : 702-709.

[10] Dammann F, Bode A, Schwaderer E, Schaich M, Heuschmid M, Maassen MM (2001) Computer-aided surgical planning for implantation of hearing aids based on CT data in a VR environment. Radiographics 21(1) : 183-191.

[11] Stieger C, Wackerlin D, Bernhard H, Stahel A, Kompis M, Hausler R et al. (2004) Computer assisted optimization of an electromagnetic transducer design for implantable hearing aids. Comput Biol Med 34(2) : 141-152.

[12] Hough JDV, Matthews P, Wood MW, Dyer RK (2002) Middle ear electromagnetic semi-implantable hearing device: Results of the Phase II SOUNDTEC direct system clinical trial. Otol Neurotol 23(6) : 895-903.

[13] Quester R, Schoder R (1997) The shrinkage of the human brain stem during formalin fixation and embedding in paraffin. J Neurosci Methods 75(1) : 81-89.

[14] Luntz M, Malatskey S, Braun J (2000) The anatomic relationship between the second genu of the facial nerve and the incus: a high-resolution computed tomography study. Am J Otol 21(5) : 686-9.

[15] Kikuchi (1903) J. Beiträge zur Anatomie des menschlichen Amboss mit Berücksichtigung der verschiedenen Rassen. [in German] Z Ohrenheilkunde 42 : 122-125.

[16] Sorensen MS, Dobrzeniecki AB, Larsen P, Frisch T, Sporring J, Darvann TA. (2002) The visible ear: a digital image library of the temporal bone. ORL J Otorhinolaryngol Relat Spec 64(6) : 378-81.

[17] Hüttenbrink KB (1992) The mechanics and function of the middle ear. Part 1: The ossicular chain and middle ear muscles. [in German] Laryngorhinootologie 71(11) : 545-551.

Chapter 4: Human temporal bones versus mechanical model to evaluate three middle ear transducers

This chapter consists of two parts. First a mechanical middle ear model which has been used to characterize the minimally invasive implantable transducer (CLT) is presented. When developing middle ear transducers, it is important to know the generated output at the stage of the cochlea. For conventional hearing aids, this is performed measuring the sound pressure level (SPL). As IHS transducers are generating forces or displacements, appropriate models are necessary to evaluate the transducers. For IHS I temporal bones are often used for characterization. This chapter shows the advantages of a mechanical middle ear model.

Then, the output of three middle ear transducers for different mounting parameters is discussed. The mechanical middle ear model is used to compare the CLT with other middle ear transducers

Reference:
Stieger C, Bernhard H, Waeckerlin D, Kompis M, Burger J, Häusler R. Human temporal bones versus mechanical model to evaluate three middle ear transducers. JRRD 2007;44:407-16.

4.1 Abstract

Three different middle ear transducers for implantable hearing aids, the driving rod transducer (DRT), the floating mass transducer (FMT or vibrant soundbridge) and the contactless transducer (CLT) were evaluated using a life-size mechanical middle ear model and human temporal bones. Results of the experiments using the mechanical model were within the range of the results for human temporal bones. However, results with the mechanical model showed better reproducibility. The handling of the mechanical model was found to be considerably simpler and less time consuming. Systematic variations of mounting parameters showed that the position and alignment on the incus have virtually no effect on the output of DRT transducer, the mass loading on the cable of the FMT has a larger impact on the output than the tightness of crimping and the output level of the CLT could be increased by 10 dB by optimizing the mounting parameters.

4.2 Introduction

Hearing disorders due to acoustic trauma, e.g. through firearm use, genetic disorders or as a part of the ageing process are frequent. Its consequences, such progressive isolation and withdrawal from social contacts are serious. It is estimated that 16% of the population in US have hearing troubles [1]. For several forms of hearing troubles, hearing aids are the preferred treatment method. Despite substantial progress in this area, conventional hearing aids still suffer from a number of drawbacks, such as feedback, limited speech recognition due to residual distortions of the loudspeaker or occlusion of the external auditory canal. Implantable hearing aids have the potential to solve these problems. Over the past decades, different implantable hearing aids have been developed [2-6] and some have even been commercialized [3, 5, 6]. In conventional hearing aids, the amplified and preprocessed sound signal is emitted by a miniature loudspeaker into the external auditory canal (EAC). The ensuing pressure variations result in vibrations of the tympanic membrane (TM), which leads to a movement of the ossicular chain of the middle ear (cf. Fig. 4.1). The sound is then transferred via the three ossicles (malleus MA, incus IN, and stapes ST) and the oval window (OW) to the cochlea, where the mechanical movement is transformed into a neural response transmitted by the auditory nerve. In contrast, implantable output transducers drive the ossicular chain directly and thereby stimulate the inner ear. Thus, implantable output transducers are the equivalent to the loudspeaker in conventional hearing aids and constitute key components of implantable hearing aids.

The evaluation of implantable transducers is an important and non-trivial part of the design an validation process. Output levels and frequency response have been measured and reported using different ear models. In simple, mechanical, non-anatomical models [7] , the entire middle ear structure is missing and, therefore, the possibilities to study the impact of different mounting parameters are fundamentally limited. In contrast, human temporal bones [6-9] in vivo animals [8, 10, 11] or in vivo human subjects [12, 13] are much better in terms of anatomy or estimation of the output levels obtained [14, 15]. However, the handling of such biological models is delicate and time consuming (15). Furthermore, difficulties arise for studies involving systematic variations of one or more mounting parameters as individual differences of more than 10 dB between temporal bones exist [15-17]. Finally, reuse and conservation time are limited [18]. A life-size mechanical middle ear model could reduce these drawbacks substantially and allow easier systematic measurements of different transducer designs.

The aim of this study is twofold. The primary aim was to assess the influence of the most important mounting parameters of three different types of implantable hearing aid transducers by way of systematic experimental variations. For this part of the study, a life-size mechanical middle ear model [19] was used. As a necessary precondition, this middle ear model had to be validated by comparative measurements in real human temporal bones.

4.3 Methods

4.3.1 Transducers

Three different electromagnetic transducers were investigated: a driving rod transducer (DRT), a contactless transducer (CLT), and a floating mass transducer (FMT) (Fig. 4.1). All transducers stimulated the inner ear by applying a force to the the incus. The resulting vibrations were then conducted via the stapes and oval window to the inner ear.

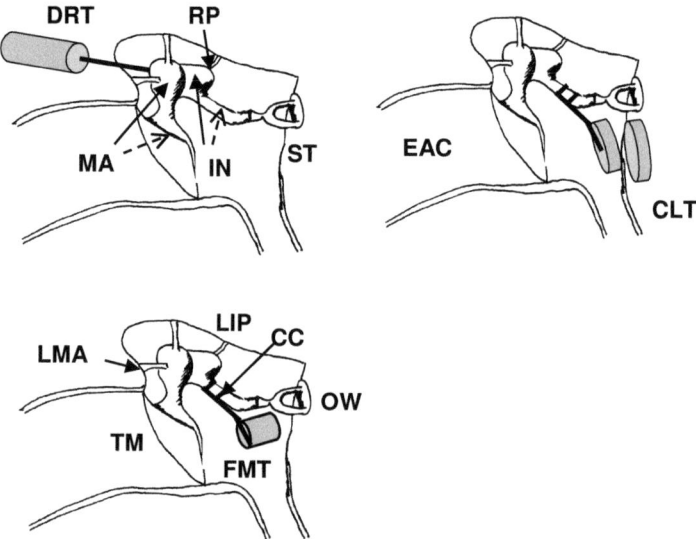

Figure 4.1: Schematic view of the three types of transducers: a) Driving rod transducer (DRT) coupled to the incus (IN) via a rod. b) Contactless transducer (CLT) consisting of a coil attached to the incus (IN) and permanent magnet attached to the wall of the middle ear cavity. c): floating mass transducer (FMT) attached to the incus (IN) via a crimp connection (CC) (EAC: external auditory canal, TM: tympanic membrane, IN: incus (short process: solid line, lenticular process: dotted line), MA: malleus (malleus head: solid line, manubrium mallei: dotted line), ST: stapes, OW: oval window, ligamentum incudis posterior

(LIP) and ligamentum mallei anterior (LMA), RP: reference point for systematic variations of the DRT coupling, cf. text.

The driving rod transducer (DRT) (Fig. 4.1a) was a custom-made device based on a similar principle as the commercially available MET™ (Otologics, Inc., Boulder, CO, USA) transducer [5] or the older totally implantable cochlea amplifier (TICA) [16]. The DRT generated a force within a hermetically sealed casing which was fixed to the mastoid (bony structure behind the external auditory canal). Its output force was applied directly to one of the ossicles by means of a coupling rod (Fig 1a).

The contactless transducer (CLT) (Fig. 4.1b) was based on a minimally invasive implantable electromagnetic transducer design, described in detail by Stieger et al. 2004 [20]. It consisted of a miniature disc-shaped coil (outer diameter 4.2 mm, length 0.3 mm) and a permanent magnet (SmCo, diameter 3.2 mm, length 0.3 mm, axial magnetization). The coil was attached to the wall of the middle ear cavity and the magnet is fixed to the long process of the incus. A current applied to the coil induced a force on the magnet which vibrated the stapes. This transducer combined the advantages of two previously presented contactless electromagnetic designs [3, 8]; i.e the external auditory canal remained open and a minimally invasive surgical technique.

The floating mass transducer (FMT) (Fig. 4.1c) was a component of the commercially available Vibrant Soundbridge implantable hearing aid system (Vibrant Med-El, Innsbruck, Austria). It consisted of a moving permanent magnet inside of a coil. Because of the inertia of the mass of the tiny magnet, the coil vibrated when an AC current was applied to it. The FMT was attached to the incus, where it vibrated the ossicular chain.

4.3.2 Mechanical middle ear model

A physical, life-sized mechanical middle ear model first presented by Taschke et al. [19] was used. This model consisted of three synthetic ossicles (malleus, incus and stapes), a tympanic membrane and three ligaments (ligamentum incudis posterior, ligamentum mallei anterior, annular ligament). and were mounted on a base plate (Fig. 4.2).

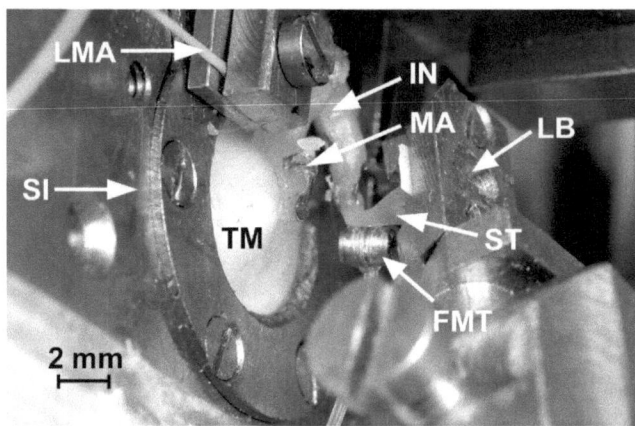

Figure 4.2: Life-size mechanical middle ear model including tympanic membrane (TM), malleus (MA), incus (IN), stapes (ST), ligamentum mallei anterior (LMA), ligamentum incudis posterior (not visible). A FMT is attached to the incus (IN). A laserbeam (LB) was directed to the backside of the footplate of the stapes (ST). Sound input (SI) was applied through an acoustically sealed chamber on the reverse side of the tympanic membrane (TM).

The tympanic membrane was made of 0.1 mm silicon rubber. It was damped with a thin layer of vaseline on the pars flaccida. The ossicles were made of a composite of epoxy resin and barite and had the average shape, size and weight of human ossicles. The incudo-stapedial joint was modeled with a droplet of latex. As a consequence both the incus and stapes could counter-shift in directions parallel to the stapes footplate in accordance with the modes of vibration at the incudo-stapedial joint in the natural human middle ear [18]. The incudo-mallear joint was rigid. The annular ligament was represented by a thin foil made of silicon rubber (0.1 mm). The foil was clamped between two brass plates, both containing a hole with the dimensions of the oval window in the human ear. The stapedial footplate was attached to the foil in this artificial oval window. A sound chamber (2 ml) corresponding to the external auditory canal was mounted in front of the tympanic membrane.

The coil of the CLT and the driver of the DRT were attached to a small custom-built mounting unit. The permanent magnet of the contactless transducer and the FMT were directly crimped to the model incus and, in addition, fixed with a removable adhesive (Crystalbond™ 555, Aremco Products Inc, NY, US). The coupling rod of the DRT was attached to the ossicles with Crystalbond™ adhesive.

4.3.3 Human cadaver temporal bones

Four fresh human temporal bones were harvested and preserved using a 1:10,000 merthiolate solution as described by Heiland et al. [18]. To perform acoustical measurements with a laser vibrometer, the laser beam was focused on the stapedial footplate. This allowed visual access to the middle ear cavity, which was obtained by means of a mastoidectomy; a common, routine, surgical approach [18]. A 2 ml sound chamber was attached and fixed with cement (TempBond, Kerr Co., Orange, CA, USA) to the bony wall of the external auditory canal.

For all temporal bones, measurements were first performed without an implantable transducer and then with an implanted transducer. For the DRT, there was no visible difference in the placement in each of the temporal bones used. For the CLT, one implantation was performed with 0 mm offset between the axis of the coil and the permanent magnet and one with an offset of 2 mm to account for anatomic variations.

The preparation for the acoustical measurement took approximately two hours and the implantation of the transducers an additional 1 to 2 hours.

All acoustical and transducer output measurements were performed in a climatic chamber at temperatures of 36° +/- 2° C and a relative humidity above 99% to prevent dehydration of the temporal bone [21, 22].

The use of human temporal bones and the study protocol were approved by the local ethical committee.

4.3.4 Equipment

Laser Doppler vibrometry is a standard measurement method in middle ear and transducer research [16, 18, 22-24] and was used to measure the displacement of the stapedial footplate (Fig. 4.3) in these experiments.

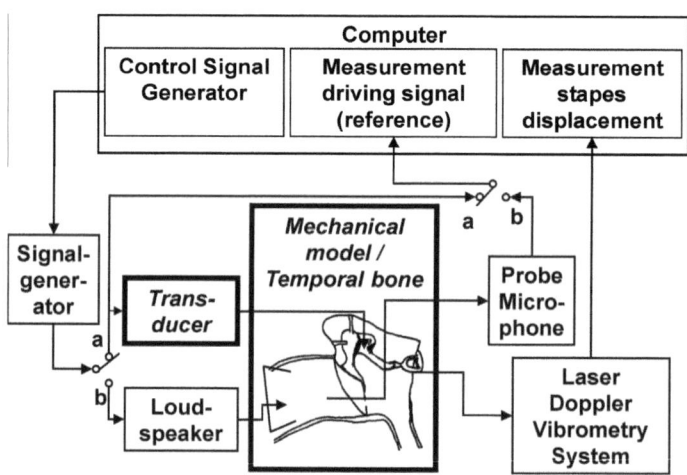

Figure 4.3: Setup for measurements on temporal bones and mechanical middle ear. A signal generator drove either one of the three transducers (a) or a loudspeaker (b). In the second case, the reference signal was given by the sound pressure level measured by a probe microphone. The displacement of the stapes was measured by laser Doppler vibrometry.

Two different types of measurements, acoustic excitation (marked (b) in figure 4.3) and transducer output measurements ((a) in Figure 4.3) were performed.

CLT evaluation

For acoustic excitation, a standard setup [14, 18] with a loudspeaker and a probe microphone (ER-7, Etymotic Research, Elk Grove Village, IL, USA) in an acoustic chamber connected to the external auditory canal was used. The acoustical characteristic was defined as the transfer function between an acoustical input at 90 dB SPL (sound pressure level) at the external auditory canal and the stapes displacement as measured by laser Doppler vibrometry.

For the transducer output measurement in the model ear and in the human temporal bones, an electrical signal was applied directly to the transducer. The transducer output measurement was defined as the transfer function between an electrical input of 1 mW of the transducer and the stapes displacement as measured by laser Doppler vibrometry.

The stimulus signal was a sinus sweep between 100 Hz and 10000 Hz. Signal analysis was performed in 21 consecutive third-octave bands within this range. Reference signal levels were 90 dB SPL at the probe microphone for acoustic excitation (acoustical measurements) and 1 mW electrical input for the transducer output measurements.

4.3.5 Comparison of the mechanical middle ear model vs. human temporal bones

In a first step the mechanical middle ear model was compared with human temporal bones. This included comparing results with acoustical excitation (Fig 4.3a) and results on transducer output measurements (Fig 4.3b) from literature and from our own measurements.

The FMT transducer was tested in the mechanical middle ear model and results were directly compared with human temporal bone measurements previously reported by Winter [13]. For the DRT and CLT transducers, temporal bone measurements were performed, as no such data is available in the scientific literature.

The acoustical measurements in human temporal bones were always performed before implantation of the transducer in order to exclude effects of possible alterations due to the explantantion of the transducers. For the mechanical middle ear, the transfer functions were measured at the beginning and the end of each session.

4.3.6 Systematic variation of transducer mounting parameters

For the systematic variations of mounting parameters the mechanical middle ear model was used.

For the DRT two parameters were varied : (1) the contact point position of the rod on the short process on the incus (tip of short process labelled RP in Figure 1a) and (2) the angle between the rod and the axis of the stapedial footplate.

For the CLT three parameters were varied: (1) the width of the air gap between coil and magnet, (2) the offset between the axis of the coil and the magnet, and (3) angle between the axes of the coil and the permanent magnet.

For the FMT two parameters were varied: (1) the use or non-use of adhesive additionally to crimping and (2) the presence or absence of a moderate mass loading (150 mg) on the lead to the transducer at a distance of 4 mm from the FMT.

The ranges of all mounting parameters were chosen to represent realistic anatomical and surgical variations [25]. The time taken for exchanging a transducer was approximately 20 minutes. Changing a single mounting parameter took approximately 5 minutes.

4.4 Results

Comparative measurements in mechanical middle ear model vs. human temporal bones

Figure 4.4 shows stapes displacements as a function of frequency for acoustical stimulations at 90 dB SPL in all temporal bones and in the mechanical model at the beginning and the end of the study.

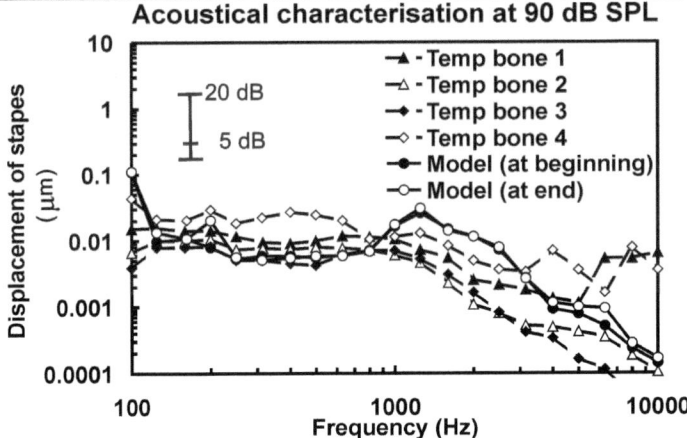

Figure 4.4: Displacement of the stapes for an acoustical excitation of 90 dB SPL

All measurements produced a plateau from 100 Hz up to 800 Hz and then decreased by approximately 40 dB per decade for higher frequencies. Stapes displacements in the mechanical ear model were larger for the higher frequencies above 1000 Hz, than with the temporal bones.

For the mechanical middle ear model, both sets of measurements were virtually identical (mean difference 1.0 dB). In 18 out of 21 third-octave bands, displacements differ by less than 2 dB.

In contrast, variations between the temporal bone measurements are significantly larger than those of the mechanical middle ear model (range from 10 dB at 1250 Hz to 55 dB at 10000 Hz).

Figure 4.5 shows the comparison between the mechanical middle ear model and the temporal bones for all transducers. The transducer output measurement of the driving rod transducer DRT was shown in Figure 5a. All measurements showed the same basic characteristic, similar to the acoustical model, i.e. they were flat up to approximately 500 Hz, then decreased with a slope of about 40 dB per decade for frequencies above 1000 Hz. They featured a resonance peak of approximately 10 dB around 800 Hz. Displacements for the temporal bones were slightly higher than for the mechanical middle ear model for frequencies below 800 Hz (mean difference 7 dB) and virtually identical (mean difference below 1 dB) for frequencies between 1250 Hz and 5000 Hz. The mean difference between the two separate measurements in the mechanical middle ear model was below 1 dB. The difference between the two temporal bones is slightly larger (mean 4.8 dB).

Figures 4.5b and 4.5d show the transducer output measurements for the contactless transducer CLT. For the first set of measurements (Figure 4.5b), the axes of the coil and the magnet were coincident (0 mm offset). For both, the temporal bone and the mechanical middle ear model, amplitudes decrease by approximately 40 dB between 250 Hz and 8000 Hz. However, between 1600 Hz and 6300 Hz, the frequency response of the mechanical middle ear model was somewhat higher than the temporal bone. For the entire frequency range, the mean difference between the two measurements in the mechanical middle ear model was below 2 dB.

The second set of measurements (Figure 4.5d) refers to the CLT with an offset of 2 mm between the axis of the coil and permanent magnet. The transducer output in the temporal bone and the mechanical middle ear model showed a plateau at around 0.1 µm for frequencies below 800 Hz. Between 800 Hz and 10000 Hz, the amplitude dropped by 40 dB, similar to the situation at 0 mm offset. Again, in the mechanical middle ear model amplitudes were somewhat higher in the range between 800 Hz and 8000 Hz. The mean difference between the mechanical middle ear measurements was small (1.2 dB).

Figure 4.5c shows the transducer output measurement for the floating mass transducer FMT. The amplitudes increased for frequencies below 800 by approximately 50 dB per decade, with a wide peak around 800 Hz to 1600 Hz, and decreased by approximately 50 dB for higher frequencies. Amplitudes for the mechanical middle ear model tended to be somewhat higher for frequencies below 1250 Hz and lower for frequencies in the range between 1250 Hz and 10000 Hz compared to the temporal bone measurements, leading to a mean difference of 0.7 dB

CLT evaluation 77

over the entire frequency range. Mean differences between the mechanical middle ear measurements were also 0.7 dB.

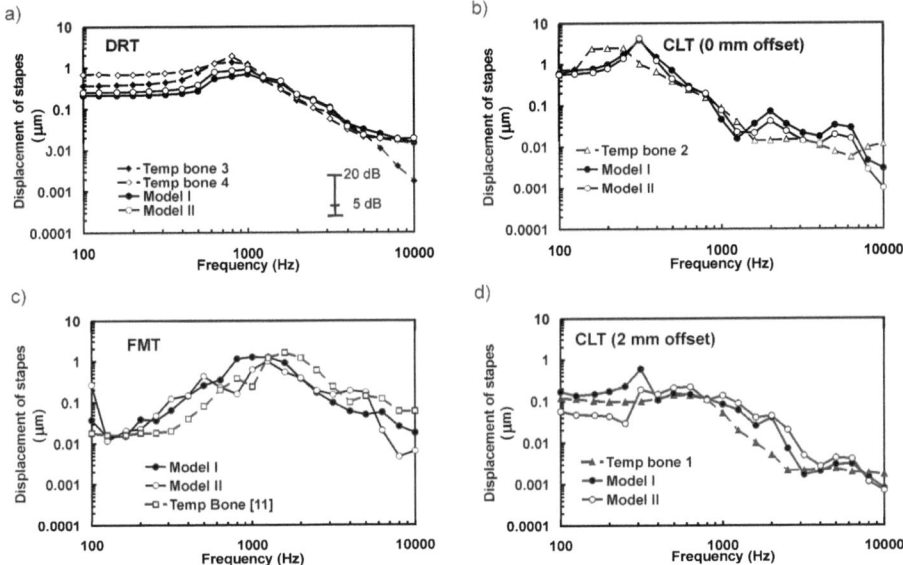

Figure 4.5: Comparison of measurements in temporal bones and in the life-size mechanical middle ear model for three transducers. a) Driving rod transducer (DRT), b,d) Contactless transducer (CLT) c) Floating mass transducer (FMT)

4.4.1 Systematic variation of transducer mounting parameters

Systematic variations of transducer mounting parameters have been performed in the mechanical middle ear model. Figure 4.6 show the results of the transducer output measurement of the driving rod transducer (DRT) for systematical variation of two mounting parameters. Figure 6a shows that the output varied by no more than 10 dB if the coupling point on the short process of the incus was varied from 0.5 mm to 2 mm, measured from the reference point (tip of the short process of the incus, labeled RP in Figure 4.1). Figure 6b shows that the output remained within a range of approximately 6 dB if the angle between the rod and the perpendicular axis of the stapes was widened from 7° to 34°.

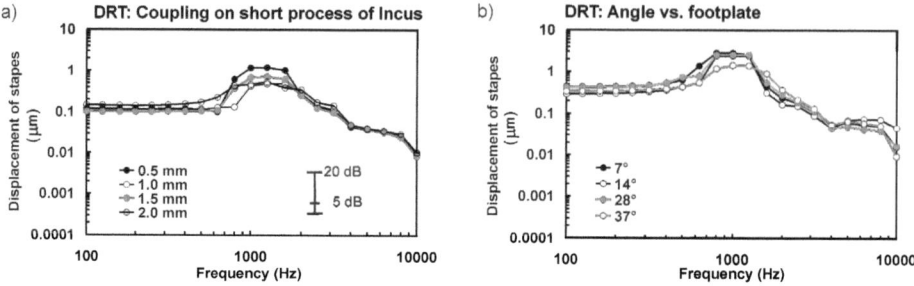

Figure 4.6: *Influence of two mounting parameters on the output of the driving rod transducer (DRT) measured in the mechanical middle ear model at 1 mW input. Coupling position on the short process of the incus (a) and angle versus footplate (b) are varied.*

Figure 4.7 shows the transducer output measurement of the contactless transducer (CLT) for systematic variation of air gap, offset and angle between the axes of the coil and the permanent magnet. If the air gap between coil and magnet was widened from 0.2 to 1.2 mm, the output decreased uniformly over the entire frequency range by an average of 10.5 dB (Fig. 4.7a). If the offset between the axes of the coil and the permanent magnet was increased moderately from 0 to 1.0 mm, the average amplitude over the entire frequency range decreased by no more than 4.6 dB. However, for larger offsets, the characteristics of the curve changed considerably. At 1.6 mm and 2.1 mm offset, amplitudes decreased predominately below 500 Hz by approximately 8 dB and above 2000 Hz by approximately 12 dB with a much smaller decreased between these two frequencies. For an offset of 2.5 mm, the amplitude decreased by approximately 29.0 dB over the entire frequency range (Fig 4.7b). Figure 4.7c shows that the maximal amplitude difference for variations of angles between the axes of the coil and the permanent magnet in the range of 0°-20° was

CLT evaluation

2.6 dB (average over the entire frequency range). At 23°, a contact between coil and magnet occured, turning the transducer into a highly nonlinear system.

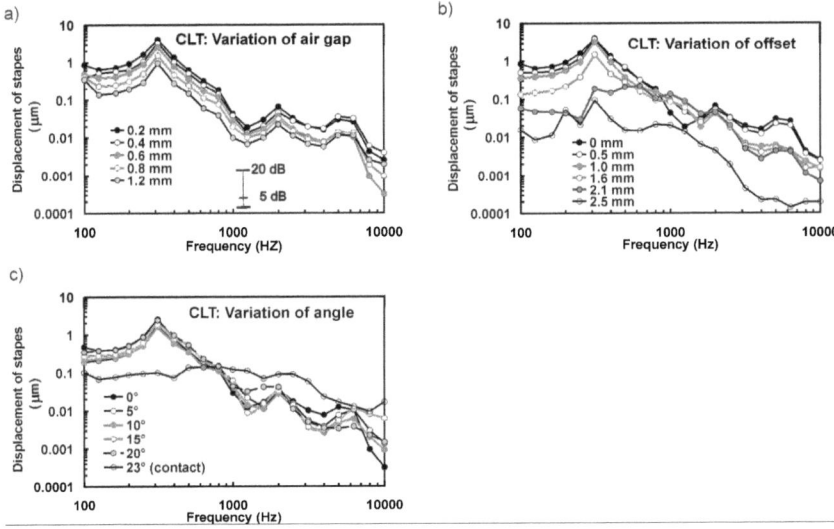

Figure 4.7: Influence of three mounting parameters on the output of the contactless transducer (CLT) measured in the mechanical middle ear model at 1 mW input. Air gap (a), offset (b) and angle (c) between coil and magnet were varied.

Figure 4.8 shows the transducer output for the FMT for moderate crimping (i.e. crimping alone) and for tight crimping (i.e. crimping plus additional fixation with CrystalbondTM adhesive), when the cable from the transducer was hanging freely (no load) and being loaded with 150 mg. Measurements for tight and moderate crimping with no load were virtually identical. For the loaded conditions, the amplitude was reduced between 700 Hz and 1500 Hz by up to 10.0 dB for tight crimping and 18.9 dB for moderate crimping. For moderate crimping with load, another output reduction of up to 26.3 dB was observed at frequencies above 5000 Hz.

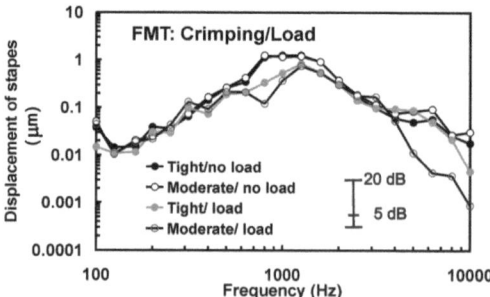

Figure 4.8: Influence of two mounting parameters on the output of the floating mass transducer (FMT) measured in the mechanical middle ear model at 1 mW input. Load on the lead and crimping are varied.

4.5 Discussion

The main purpose of this study was to examine the influence of different mounting parameters on the output of three implantable middle ear transducers, namely a driving rod transducer (DRT), a contactless transducer (CLT) and a floating mass transducer (FMT). A total of seven mounting parameters for these three transducers were measured in a life-sized mechanical middle ear model. As a prelude and a prerequisite, measurements in the mechanical middle ear model were compared with real human temporal bones measurements for each of the three transducers.

The output of the DRT was almost invariant for the examined parameters in this study. The exact positioning on the short process of the incus and the angle could be varied in a wide range with virtually no influence on the transducer output. The output will be therefore hardly influenced by individual alignment of implantation.

For the CLT, the output could be increased by up 10 dB by reducing the air gap from 1.2 mm to 0.2 mm. If coil and magnet were not coplanar, the angle between the two components had virtually no influence on the output level, as long as the magnet and the coil were not touching (Fig. 4.7c). The influence of touching was highly non linear. As a consequence, an optimal surgical placement (i.e. a small gap) is delicate because of the danger of direct contact. For practical applications, slightly larger gaps might therefore be preferable.

Small offsets between the axes of coil and magnet of up to 1 mm, i.e until the axis of the magnet was directly coincident with the brim of the coil, have little influence on the output. However, larger offsets (i.e. offset is larger than the radius of the permanent magnet) should be avoided as they reduce the low frequency output significantly. This may be due to the fact that for large offsets, the component of the horizontal force parallel to the plane of the magnet and coil becomes dominant [20] and therefore a reduction of output force in the perpendicular direction occurs. Results of the mechanical middle ear model confirmed results of a computer model [20] regarding the dependence of the expected output on the width of the air gap. However, this computer model alone could not predict the order of magnitude of non-linearity when the two components of the system touch

For the output of the FMT, the quality of crimping seemed to have virtually no influence on the acoustical output, which corresponds to the results of Snik and Cremers [12]. They reported no significant changes in 5 out of 6 frequencies between 250 Hz and 6000 Hz if cement was used for additional fixation.

According to our results, the load on the electrical cable has more impact on the output of FMT than using additional adhesive for the crimping.

The life-sized mechanical middle ear model [19] was found to be a useful instrument to examine the mounting parameters of the transducers used for this study as it allows systematic variations of single parameters. Preparation time for a single set of measurements was dramatically shorter than for human temporal bones. This is mainly due to the necessity of extensive drilling and the more time consuming mounting of the transducers in a much more complex and limiting real anatomical environment. In contrast to human temporal bones, the mechanical middle ear model was available at any time and virtually for any duration Furthermore, the mechanical middle ear model was found to be stable over extended periods of time, and repeated measurements were reproducible, confirming the results of Taschke et al. [19] even for measurements with implantable transducers. In contrast, human temporal bones can only be used for a limited time[18], even when frozen and [24] interindividual variations in the order of magnitude of 10 dB occurred in agreement with previous reports [15-17, 22].

For frequencies above 1000 Hz for direct acoustical stimulation (Fig. 3b) displacements were approximately 7 dB higher for the mechanical middle ear model than for human temporal bones. This result is in agreement with the expected transmission loss in the incudo-mallear joint for frequencies above 1000 Hz [26]. In this study all investigated transducers have been coupled to the incus, i.e. after the incudo-mallear joint. Therefore, the performance of transducer output measurements in the mechanical model are not expected to be affected substantially by the rigidity of the incudo-mallear joint. This has been confirmed by our measurements.

Our results allow the direct comparison between different transducer designs (Fig. 4.5). In this comparison, the FMT transducer generates smaller displacements at low frequencies below 800 Hz which corresponds to one limitation of the current audiological indication [27]. The CLT generates lower displacements than the other two transducer types in the frequency range between 1000-3000 Hz. The DRT tends to be the most efficient transducer almost over the entire frequency range.

4.6 Conclusions

Implantable hearing aid transducers can be evaluated using the life-size mechanical middle ear model [19]. Compared to tests using human temporal bones, the handling was found to be simpler and less time consuming. Furthermore, individual parameters could be varied systematically more easily than with human temporal bones. Results from the mechanical middle ear model and from human temporal bones were found to be in reasonable agreement with all three transducer designs considered in this research.

Regarding systematic variations of the mounting parameters of the three different middle ear transducers, it was found that the exact position of the coupling point of the DRT on the short process of the incus and the mounting angle are of minor importance for the level and frequency response of the transducer. For the output of the FMT, the quality of crimping was found to have little influence on the acoustical output, whereas the load on the cable to the transducer had a greater impact. The CLT system was found to become non-linear for large offsets and contact between the components.

4.7 References

[1] Lethbridge-Çejku M, Rose D, Vickerie J. Summary health statistics for U.S. Adults: National Health Interview Survey, 2004. National Center for Health Statistics. Vital Health Stat 2006;10(228):43.

[2] Chen DA, Backous DD, Arriaga MA, Garvin R, Kobylek D, Littman T, et al. Phase 1 clinical trial results of the Envoy System: a totally implantable middle ear device for sensorineural hearing loss. Otolaryngol Head Neck Surg 2004;131(6):904-16.

[3] Hough JV, Dyer RK, Jr., Matthews P, Wood MW. Semi-implantable electromagnetic middle ear hearing device for moderate to severe sensorineural hearing loss. Otolaryngol Clin North Am 2001;34(2):401-16.

[4] Huttenbrink KB. Current status and critical reflections on implantable hearing aids. Am J Otol 1999;20(4):409-15.

[5] Kasic JF, Fredrickson JM. The Otologics MET ossicular stimulator. Otolaryngol Clin North Am 2001;34(2):501-13.

[6] Ball, inventor Symphonix Device, assignee. Implantable and external hearing systems having a floating mass transducer. United States. 1999.

[7] Leysieffer H, Baumann JW, Muller G, Zenner HP. [An implantable piezoelectric hearing aid transducer for inner ear deafness. II: Clinical implant]. Hno 1997;45(10):801-15.

[8] Ko WH, Zhu WL, Kane M, Maniglia AJ. Engineering principles applied to implantable otologic devices. Otolaryngol Clin North Am 2001;34(2):299-314.

[9] Needham AJ, Jiang D, Bibas A, Jeronimidis G, O'Connor AF. The effects of mass loading the ossicles with a floating mass transducer on middle ear transfer function. Otol Neurotol 2005;26(2):218-24.

[10] Javel E, Grant IL, Kroll K. In vivo characterization of piezoelectric transducers for implantable hearing AIDS. Otol Neurotol 2003;24(5):784-95.

[11] Plinkert PK, Baumann JW, Lenarz T, Keiner S, Leysieffer H, Zenner HP. [In vivo studies of a piezoelectric implantable hearing aid transducer in the cat]. Hno 1997;45(10):828-39.

[12] Snik A, Cremers C. Audiometric evaluation of an attempt to optimize the fixation of the transducer of a middle-ear implant to the ossicular chain with bone cement. Clin Otolaryngol Allied Sci 2004;29(1):5-9.

[13] Winter M, Weber BP, Lenarz T. Measurement method for the assessment of transmission properties of implantable hearing aids. Biomed Tech (Berl) 2002;47 Suppl 1 Pt 2:726-7.

[14] Gan RZ, Wood MW, Ball GR, Dietz TG, Dormer KJ. Implantable hearing device performance measured by laser Doppler interferometry. Ear Nose Throat J 1997;76(5):297-9, 302, 305-9.

[15] Huber A, Linder T, Ferrazzini M, Schmid S, Dillier N, Stoeckli S, et al. Intraoperative assessment of stapes movement. Ann Otol Rhinol Laryngol 2001;110(1):31-5.

[16] Leysieffer H. [Principle requirements for an electromechanical transducer for implantable hearing aids in inner ear hearing loss. I: Technical and audiologic aspects]. Hno 1997;45(10):775-86.

[17] Nishihara S, Goode R. Measurements of tympanic membrane vibrations in 99 human ears. In: Huttenbrink KB, editor. Workshop on middle ear mechanics in research and otosurgery; 1996 Sept 19-22; Dresden: Dep. of Oto-Rhino-Laryngology,Univ. Hospital Carl Gustav Carus, Univ. of Technology; 1996. p. 91 - 95.

[18] Heiland KE, Goode RL, Asai M, Huber AM. A human temporal bone study of stapes footplate movement. Am J Otol 1999;20(1):81-6.

[19] Taschke H, Weistenhoefer C, Hudde H. A Full-Size Physical Model of the Human Middle Ear. Acustica/acta acustica 2000;86:103-116.

[20] Stieger C, Wackerlin D, Bernhard H, Stahel A, Kompis M, Hausler R, et al. Computer assisted optimization of an electromagnetic transducer design for implantable hearing aids. Computers in Biology and Medicine 2004;34(2):141-152.

[21] Eiber A, Kauf A, Maassen MM, Burkhardt C, Rodriguez J, Zenner HP. [First comparisons with laser vibrometry measurements and computer simulation of ear ossicle movements]. Hno 1997;45(7):538-44.

[22] Voss SE, Rosowski JJ, Merchant SN, Peake WT. Acoustic responses of the human middle ear. Hear Res 2000;150(1-2):43-69.

[23] Stenfelt S, Hato N, Goode RL. Factors contributing to bone conduction: the middle ear. J Acoust Soc Am 2002;111(2):947-59.

[24] Vlaming MS, Feenstra L. Studies on the mechanics of the reconstructed human middle ear. Clin Otolaryngol Allied Sci 1986;11(6):411-22.

[25] Stieger C, Djeric D, Kompis M, Remonda L, Hausler R. Anatomical study of the human middle ear for the design of implantable hearing aids. Auris Nasus Larynx 2006.

[26] Willi UB, Ferrazzini MA, Huber AM. The incudo-malleolar joint and sound transmission losses. Hear Res 2002;174(1-2):32-44.

[27] Fraysse B, Lavieille JP, Schmerber S, Enee V, Truy E, Vincent C, et al. A multicenter study of the Vibrant Soundbridge middle ear implant: early clinical results and experience. Otol Neurotol 2001;22(6):952-61.

Chapter 5: A novel implantable hearing system with direct acoustical cochlear stimulation (DACS)

This chapter presents the concept of the DACS system. The implantation procedure is described. Finally results of the first clinical study are then provided. They show that patients with severe combined hearing loss can be effectively treated with this new concept.

This chapter is the result of a strong collaboration betweem the university Department of ENT at the Inselhospital, the hearing industry, the cochlear implant industry and microtechnology.

Reference:

Hausler R, Stieger C, Bernhard H, M. Kompis. A Novel Implantable Hearing System with Direct Acoustic Cochlear Stimulation. Audiol Neurootol 2008;13:247-56.

5.1 Abstract

A new implantable hearing system, the direct acoustic cochlear stimulator (DACS) is presented. This system is based on the principle of a power-driven stapes prosthesis and intended for the treatment of severe mixed hearing loss due to advanced otosclerosis. It consists of an implantable electromagnetic transducer, which transfers acoustic energy directly to the inner ear, and an audio processor worn externally behind the implanted ear.

The device is implanted using a specially developed retromeatal microsurgical approach. After removal of the stapes, a conventional stapes prosthesis is attached to the transducer and placed in the open window to allow direct acoustical coupling to the perilymph of the inner ear. In order to restore the natural sound transmission of the ossicular chain, a second stapes prosthesis is placed in parallel to the first one into the oval window and attached to the patient's own incus, as in a conventional stapedectomy.

Four patients were implanted with an investigational DACS device. The hearing threshold of the implanted ears before implantation ranged from 78 to 101 dB (Air conduction, pure tone average PTA, 0.5 – 4 kHz) with air-bone-gaps of 33 to 44 dB in the same frequency range.

Postoperatively, substantial improvements in sound field thresholds, speech intelligibility as well as in the subjective assessment of everyday situations were found in all patients. Two years after the implantations, monosyllabic word recognition scores in quiet at 75 dB improved by 45 to 100 percent points when using the DACS. Furthermore, hearing thresholds were already improved by the second stapes prosthesis alone by 14 to 28 dB (PTA 0.5 - 4 kHz, DACS switched off). No device related serious medical complications occurred and all patients have continued to use their device on a daily basis for over two years.

5.2 Introduction

Active hearing implants are a dynamic area of research [1-3]. When compared to conventional hearing aids, implantable aids hold the promise of substantial improvements regarding sound quality, speech recognition, sound distortion, reduced feed-back and less discomfort due to absence of ear canal occlusion [4-6].

The single most important component of an implantable hearing aid is the transducer, i.e. the equivalent of the loudspeaker in conventional hearing aids, providing a direct mechanical interface to the human ear, typically at the level of the ossicular chain [7]. Today, several types of implantable hearing aids are either available or have been proposed [1,8]. However, they currently provide either limited hearing gain or they may induce an additional hearing impairment [9] when the system is inactive. Furthermore, treatment of severe mixed hearing loss is the difficult part. This type of hearing problem can be caused by advanced otosclerosis. with an additional inner ear hearing impairment.

This group of patients is usually treated with conventional hearing aids, which, however, often do not offer sufficient gains. Alternatively, stapedectomy allows the treatment of the conductive component only. Although the combination of stapedectomy and conventional hearing aids further improves hearing, it will be shown in this paper that there is a single treatment resulting in better aided hearing thresholds.

We present a new transducer for an implantable hearing system, the DACS, an abbreviation for direct acoustic cochlear stimulator. In this report, the DACS device, the surgical procedure required for implantation and the outcome of the first clinical trial with an investigational device are presented.

5.3 Methods

5.3.1 Concept of the DACS System

The DACS concept is based on the principle of a power-driven stapes prosthesis. In contrast to other active hearing implants, it directly vibrates the fluid of the cochlea. Figures 5.1 through 5.3, which are discussed in detail later in this text, show different views of an investigational DACS system. An implantable transducer converts an electrical input signal into a movement of a driving rod, which couples to the inner ear fluid, e.g. at the level of the oval window, therefore bypassing all structures which may cause a conductive hearing loss. The transducer itself is driven by a fully or partially implantable signal processor unit, which provides appropriate amplification and signal processing to overcome the sensorineural component of the hearing loss.

In this way, the DACS unites two concepts of treatment of hearing impairments in one single system: mechanical amplification and established otological microsurgery.

5.3.2 Investigational device

The investigational device of the clinical trial is shown in Figure 5.1. It consists of an externally worn audio processor and an implanted part, consisting of the DACS transducer, a percutaneous plug, a fixation system and an "off the shelf" stapes prosthesis.

The external audio processor contains two microphones, a digital signal processing unit and a battery. It is based on a state of the digital art multi-channel hearing aid system Savia 211 (Phonak AG, Switzerland). It features multi-channel compression, noise and feedback canceling, and a multi microphone noise reduction system. The fitting software was specifically adapted for the DACS.

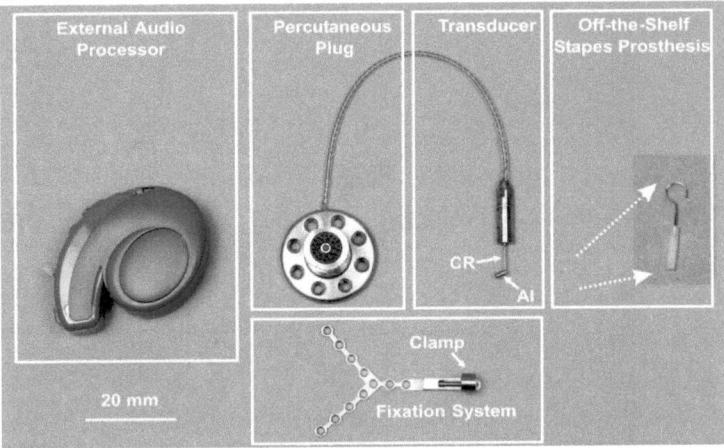

Figure 5.1: Investigational direct acoustic cochlear stimulation (DACS) device used in the pilot study. (CR: coupling rod, AI: artificial incus)

In contrast to conventional hearing aids, the electrical output of the audio processor drives the transducer by means of a percutaneous plug, which was already being used in the Ineraid cochlear implant system [10]. The transducer itself consists of a miniaturized electro-mechanical driver. A so-called balanced armature principle, which is often used for acoustic devices such as hearing aid speakers, was chosen. The employed balanced armature principle best meets the requirements considering the demanding dynamic characteristic of the human middle ear [11-13]. It provides vibration amplitudes of up to 25 μm [14] corresponding to a maximal power output of more than 125 dB SPL over the entire frequency range from 100 Hz to 10'000 Hz. Figure 5.2 shows the main functional components of the transducer: armature, magnets and coil.

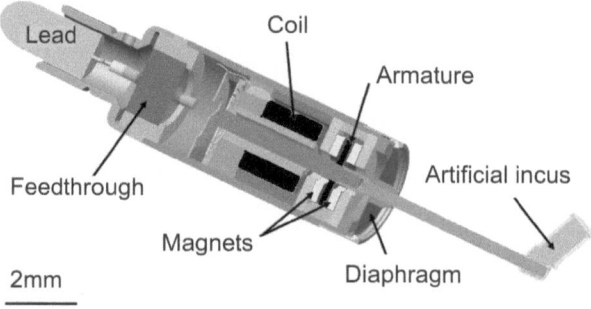

Figure 5.2: Cross section of the DACS transducer (schematic).

The transducer features a titanium diaphragm (Fig. 5.2) that allows a movable but hermetically sealed interface between armature and coupling rod [15]. The generated vibrations are transferred to the artificial incus by the means of a coupling rod of 0.4 mm diameter. The artificial incus is coated with a thin silicone layer. Its size and shape correspond to the incus long process of the human ossicular chain. A conventional stapes prosthesis can be attached by crimping in the same way as in routine stapedectomy.

All transducer parts that are in contact with human tissue are made of implantable grade materials (titanium, platinum-iridium and silicone). The non-biocompatible parts (magnets, coil and soft magnetic alloys) are hermetically encapsulated.

The fixation system that anchors the transducer to the mastoid surface of the patient (Figures 5.1 and 5.3) is based on a micro titanium bone plate - as used for cranio-maxillofacial trauma surgery - augmented with a special clamping mechanism for the DACS transducer. The fixation system can be bent by the surgeon to fit the curvature of the skull and bring the transducer into the correct position. Conventional titanium bone screws are used to fixate the plate. The fine positioning of the transducer is effected by inserting the transducer in the clamp and choosing the optimal orientation and insertion depth before closing the clamp with a torque screwdriver. The clamp is designed to be opened and closed several times, if necessary, during implantation, although it was rarely required in the implantations performed so far.

5.3.3 Surgical procedure: implantation by the "retromeatal approach"

The surgical procedure was tested and refined first using temporal bones and a total of 27 anatomical specimens of entire human cadaver heads. A detailed surgical 50 step protocol was developed interactively by surgeons and the designers of the implant. The surgical procedure is focused on patient safety first and on the optimal configuration and placement of the implant and its components as a close second.

A special "retromeatal approach" derived from a minimally invasive cochlea implantation procedure [16] was developed to place the transducer at its intended position in the mastoid bone (Fig. 5.3).

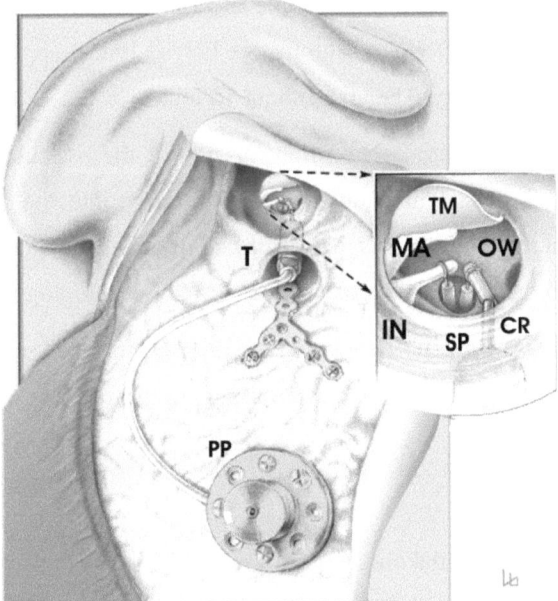

Figure 5.3: Artist's rendition of the DACS towards the end of implantation surgery (PP: percutaneous plug, T: transducer, SP: stapes prostheses, CR coupling rod, AI: artificial incus, OW: oval window, TM: tympanic

membrane, IN: incus, MA: malleus) In this approach, the electro-mechanical transducer is implanted behind the ear. After drilling a bony tunnel behind the external auditory canal down close to the facial nerve (corresponding to a small mastoidectomy), a posterior tympanotomy by facial recess approach is performed at the level of the oval window.

The transducer is then placed in the tunnel by positioning the rod close to the long process of the incus in the tympanic cavity. The otosclerotically fixed stapes is totally removed. To allow acoustical coupling of the DACS to the perilymph, a conventional, commercially available stapes prosthesis is placed in the open oval window and crimped onto the artificial incus of the transducer.

As an inherent part of the surgical 50-step procedure, the DACS is tested intra-operatively. It is required to pass a simple electrical test as well as a mechanical vibration test using laser Doppler vibrometry. For a clinical application, these tests will not be strictly necessary. They take about 15 min.

In order to restore the natural sound transmission of the ossicular chain, a second stapes prosthesis is placed in parallel to the first one into the oval window and attached to the patient's own incus, as performed in conventional stapedectomy. The oval window with the two stapes prostheses is sealed with autologous adipose tissue.

Again, the second stapes is not strictly necessary for the functionality of the DACS device itself, but it provides improved hearing for the patients even if the DACS system is turned off.

5.3.4 Study protocol

A study protocol for the initial clinical trial was established and approved by the local ethical committees of Berne, Switzerland, and Hanover, Germany.

Adult subjects with otosclerosis and a severe to profound mixed hearing loss were considered for inclusion in the study. Preoperative CT scans were performed. All were required to be experienced hearing aid users. They were implanted in the ear with the poorer hearing threshold. Standard intraoperative facial monitoring was performed in all patients. Pre- and postoperatively, medical and audiological evaluations were performed according to long-term audiological evaluation up to two years after implantation.

At every pre- and postoperative visit, pure tone audiograms, including air conduction (AC) and bone conduction (BC) thresholds and speech tests, were performed. Speech tests in quiet included the measurement of the speech reception threshold for 50% (SRT50%) speech intelligibility of German two-digit numbers (Freiburger number-test) and the measurement of monosyllabic word understanding at 60, 75 and 75 dB SPL (Freiburger monosyllables). For French speaking patients, the disyllabic and monosyllabic Fournier-tests were used, as they are regarded as largely equivalent to the above German tests [17]. To test speech intelligibility in noise, the Basler Sentence Test [18] was used. In this adaptive test, speech babble noise at 70 dB is held constant, and 15 test items are presented using presentation levels according to an adaptive algorithm to find the speech reception threshold for 50% speech understanding.

Tests were performed with earphones and in the soundfield under aided and unaided conditions. Masking of the contralateral side was when necessary. Soundfield measurements were performed under three different conditions: in condition I, the contralateral ear was occluded with an earplug (E-A-R Classic, Aearo Company, Indianapolis, USA) with a specified average attenuation between 24.6 and 41.6 dB in the range of 250 to 4000 Hz. The DACS was switched off. Condition II was the same as condition I, but with the DACS switched on. In condition III, both ears were plugged, but the DACS device was active. This last condition was included to examine the effect of potential interferences between direct sound and the output of the DACS system.

Every visit included an otoscopy and a tympanometry. A single patient visit took approximately 3 hours.

In addition, in three patients, soundfield thresholds were measured postoperatively with a conventional hearing aid, at the DACS-ear with the DACS switched off. Again, the contralateral ear was occluded. The hearing aid used had the same signal processor and fitting strategy as the sound processor of the hearing aid in order to minimize bias from different signal processing strategies [19].

An Abbreviated Profile on Hearing Aid Benefit (APHAB [20]) was completed by all subjects preoperatively and 12 months postoperatively. The APHAB questionnaire consists of 24 questions, classified into four scales. Ease of communication (EC) describes the effort in communication under relatively easy listening conditions. Reverberation (RV) describes understanding in moderately reverberant rooms. Background noise (BN) describes the speech understanding in the presence of multi-talker babble or other environmental competing noise. Aversiveness of Sound (AV) describes whether loud environmental sound is tolerated or results in negative reactions.

Patients were asked to rate their hearing under their usual every-day conditions, i.e. with the DACS and a contralateral hearing aid, if one was used, and with the DACS alone, if no contralateral hearing aid was used.

5.3.5 Subjects

Four patients, ages 35 to 71, participated in the study. Table 5.1 shows a synopsis of patient related data. All subjects agreed to participate in the study after giving their informed consent in writing. All were experienced but dissatisfied hearing aid users and all suffered from a severe to profound mixed hearing loss due to advanced otosclerosis (cf. pre-operative audiograms in figure 5.5), which was confirmed intra-operatively. The sensorineural component was 30 dB or more for all frequencies above 500 Hz and no notable progression of the hearing loss during the last 12 months was observed. The DACS was implanted in the audiologically poorer ear.

Table 5.1: Patient assessment in the pilot study

Patient No.	Gender	Age at implantation	Implanted ear	Study Center
01	Male	35	Right	Inselspital Bern
02	Female	60	Right	Inselspital Bern
03	Male	54	Left	Inselspital Bern
04	Female	71	Right	Medizinische Hochschule Hannover

5.4 Results

5.4.1 Surgery

The pre-defined surgical procedure was followed in all 4 patients, resulting in uneventful implantation at both of the centers involved. The total time of surgery was 5 hours for the first implantation and 2.5 – 3.5 hours for the next 2 surgeries. The experiences gained during the initial surgery yielded a considerably shorter implantation time for the second and third patient in Berne. The implantation of patient 04 in the second center (Medizinische Hochschule Hanover, MHH, Gemany) took approximately 4 hours.

5.4.2 Patient post-surgical recovery

All patients went through surgery without notable problems. In particular, none of the surgeries led to any additional hearing loss, additional tinnitus or facial palsy. Type A tympanograms were measured in all patients postoperatively. Figure 4 shows a postoperative photograph of patient 02 with the audio processor in situ.

Patient 01 reported some postoperative pain and transitory dizziness and a temporary dysgeusia. Patient 02 reported no problems whatsoever. Patients 03 and 04 experienced temporary inflammation of the tissue surrounding the skin perforation of the percutaneous plug during rehabilitation. These were treated successfully with antibiotics.

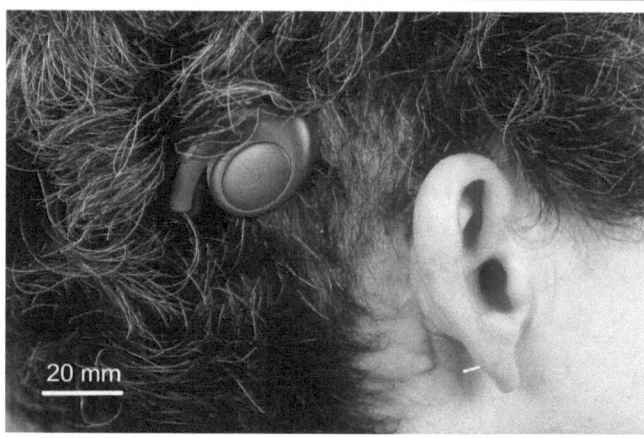

Figure 5.4: Patient 02 with the audioprocessor of the investigational device in situ.

5.4.3 Audiological outcome

Figure 5.5 shows the pure tone audiograms of all four patients. Using insert ear phones proper masking was applicable without masking dilemma [21] in all patients. Preoperative pure tone average (PTA) for the frequencies 500, 1000, 2000, and 4000 Hz for the DACS designated ear ranged between 78 to 101 dB HL (AC) with air bone gaps between 33 and 44 dB. BC thresholds ranged from 35 to 75 dB SPL in the frequency range 500 Hz to 4000 Hz. Postoperatively, unaided AC thresholds (bold in figure 5.5) were improved by 14 and 28 dB (PTA) due to the stapedectomy alone. Air bone gaps were decreased in all patients by 10 to 25 dB (PTA). BC thresholds were improved in patients 2 and 3 in the vicinity of the frequencies expected for a Carhart notch.

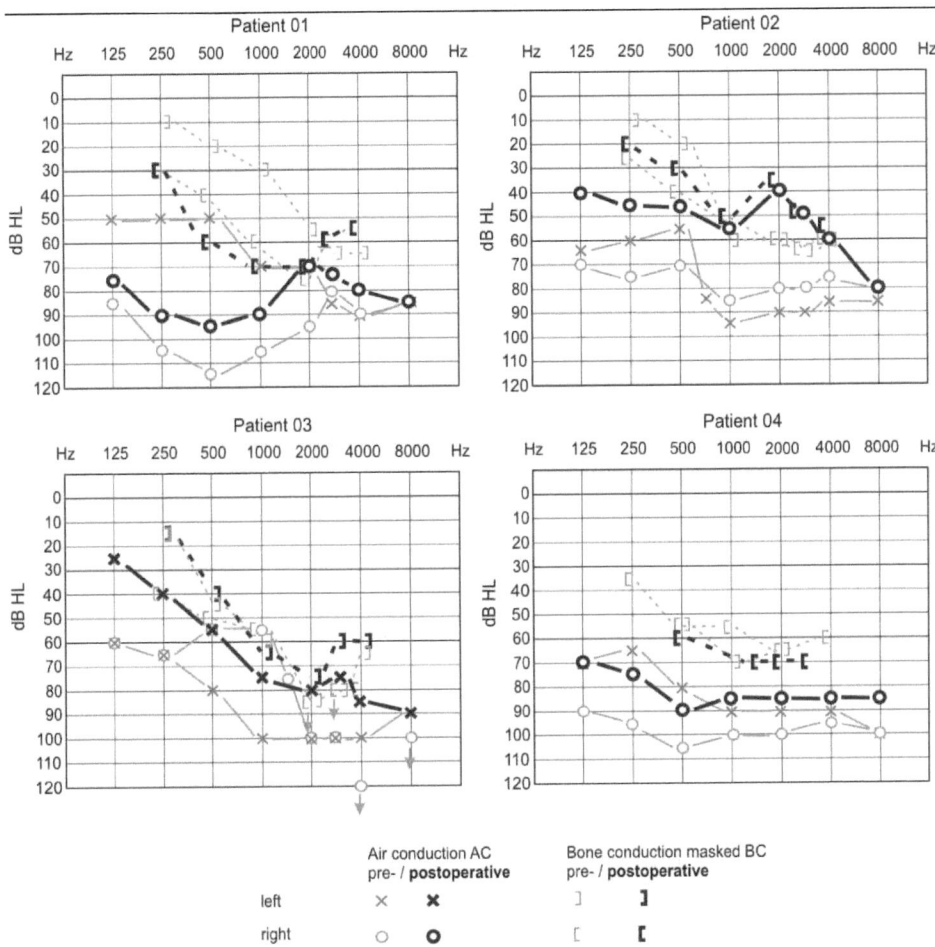

Figure 5.5: Pure tone audiograms of all patients 01-04. Bold symbols: 2-year postoperative measurements at the implanted ear. Light symbols: preoperative data.

Figure 5.6 shows the threshold measured with warble tones in the sound field with the non-implanted ear being occluded. In the unaided condition, postoperative thresholds were better than preoperative thresholds in all 4 patients. The PTA was improved by 8 to 37.5 dB. With the DACS system active, the sound field thresholds were improved by 41, 49, 50 and over 62 dB PTA (patients 01, 02, 03 and 04), respectively, when compared to preoperative measurements. Hearing thresholds were equal or better (average improvement: 7.5 dB PTA) with the activated DACS than they were with the stapedectomy and a conventional hearing aid with the same signal processing in the same ear (grey symbols in Figure 5.6).

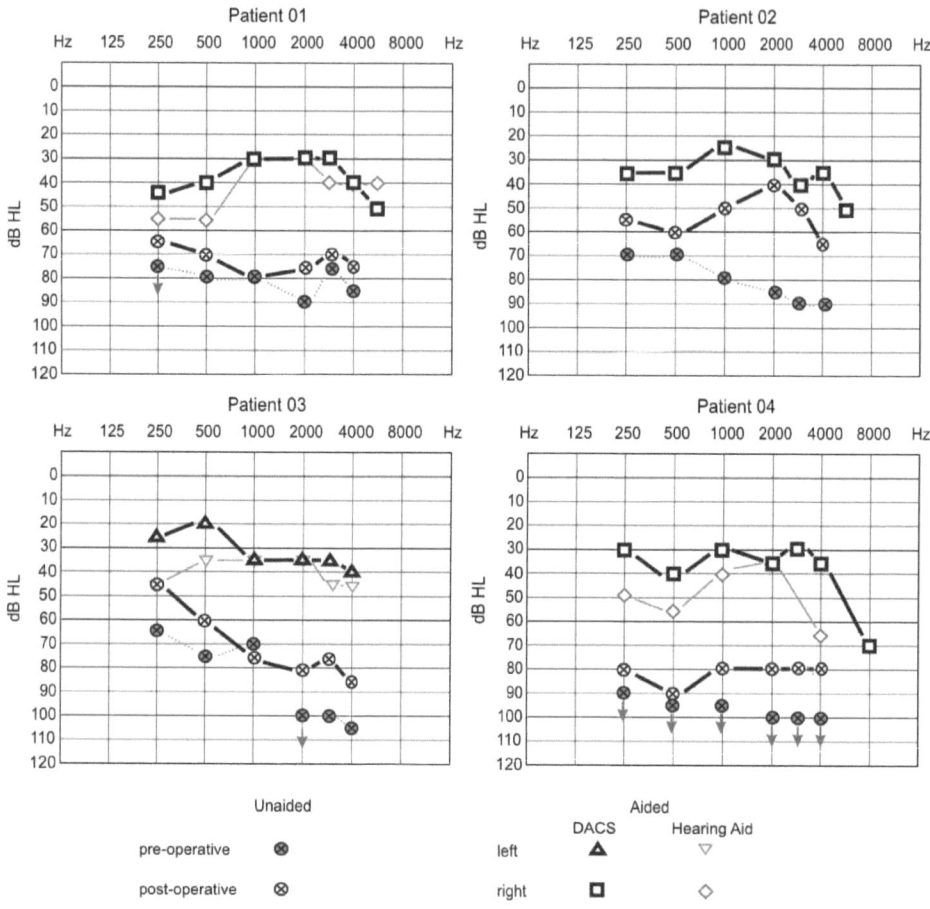

Figure 5.6: Soundfield thresholds of all patients 01-04. The hearing aid, to which the DACS was compared, featured the same signal processing capabilities and was fitted in the same implanted ear.

Table 5.2 shows the SRT50% in quiet for all patients preoperatively as well as postoperatively unaided and postoperatively with the DACS activated. SRT50% improved by 42 to 52 dB when DACS was activated and between 10-32.5 dB with the DACS switched off, due to the stapedectomy alone. Additional occlusion of the ipsilateral ear did not significantly change the SRT50%.

Table 5.2: Speech reception threshold (SRT50%) in quiet for multi-syllabic test items (numbers) in quiet of patients 01-04

SRT in dB SPL		Patient			
		01	02	03	04
Unaided	Pre OP[a]	93.5	93.5	96.5	105
	Post OP[a]	85.5	61	73.5	98
Aided	Ips open[a]	53.5	46	43.5	55
	Ips closed[b]	51.5	43.5	43.5	55

[a] ipsilateral ear was open, contralateral ear was occluded with an earplug
[b] ipsilateral ear and contralateral ear are occluded with an earplug

For the measurement of the SRT50% in noise, subjects must be able to understand 50% of the speech material [22]. As a consequence, only patients 02 and 03 were able to complete this test and postoperatively in the aided and unaided condition. The SRT50% in noise improved by 7.2 and 13.8 dB when the DACS was activated.

Table 5.3 shows speech understanding of monosyllabic words at 60, 75 and 90 dB SPL. The non-implanted ear was occluded for all measurements. Preoperatively, all patients had 0% intelligibility at all presentation levels. Postoperatively, the subjects achieved discrimination levels of 30, 70, 55 and 0% (patients 01, 02, 03, 04) when the DACS was not activated.

With the activated DACS, monosyllabic speech understanding improves for all patients at all presentation levels by 15 to 100%. Patients 02 and 03 even achieved speech recognition scores of 100 % at 75 dB. Patient 04, with the poorest performance, reached 40% at the same presentation level.

Table 5.3: Speech intelligibility for monosyllabic words.

		Presentation Level (dB SPL)	Patient			
			01	02	03	04
Unaided	Pre-op[a]	60	0 %	0 %	0 %	0 %
		75	0 %	0 %	0 %	0 %
		90	0 %	0 %	0 %	0 %
	Post-op[a]	60	0 %	10 %	0 %	0 %
		75	0 %	50 %	0 %	0 %
		90	30 %	70 %	55 %	0 %
Aided	Ips open[a]	60	25 %	80%	70 %	10 %
		75	65 %	100%	100 %	40%
		90	70 %	90%	90%	25%
	Ips closed[b]	60	20 %	80%	75 %	10 %
		75	55 %	100 %	95 %	40 %
		90	65 %	90 %	95 %	25 %

[a] ipsilateral ear was open, contralateral ear was occluded with an earplug,
[b] ipsilateral ear and contralateral ear are occluded with an earplug

Figure 5.7 shows the summary of the pre- and postoperative APHAB scores for all patients. A difference of 10% or more in any of the three subscales Ease of Communication (EC), Reverberation (RV) and Background Noise (BN) is considered a significant difference between different conditions at the 95% confidence level [20]. For the last subscale, Aversiveness of Sounds (AV), no or only a small difference is expected. Lower values denote more favorable assessments in all subscales.

DACS

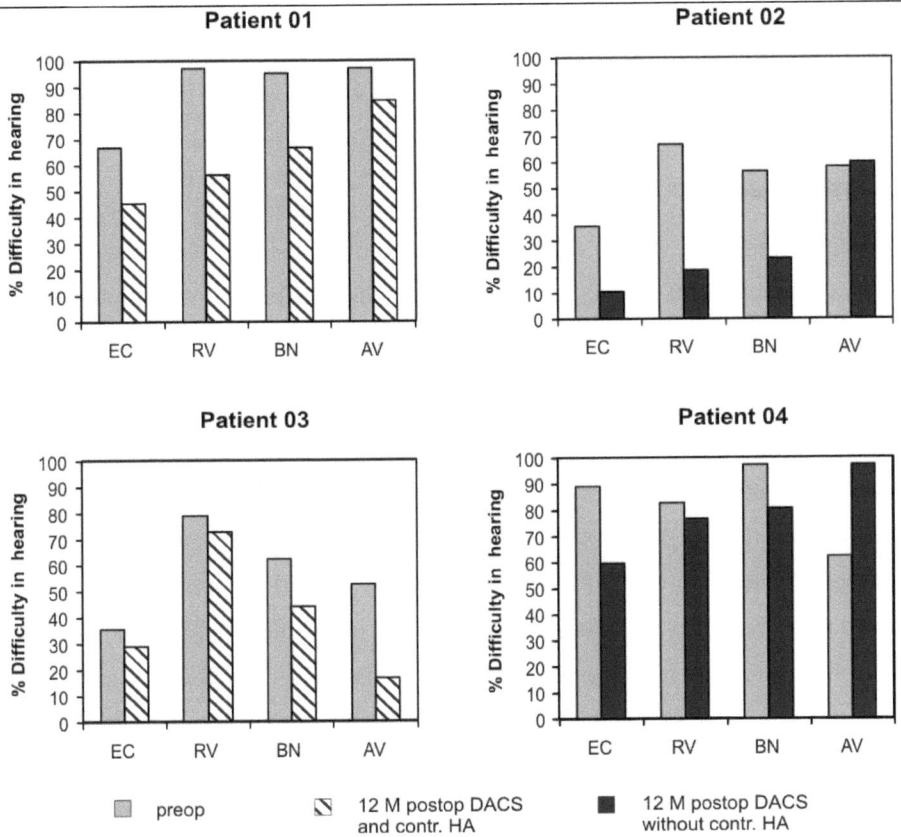

Figure 5.7: Scores of the APHAB questionnaire for patients 01-04. EC: ease of communication, RV: reverberation, BN: background noise, AV: aversiveness); lower scores denote a better assessment by the users.

Patient 01 reported improvement of communication in all 4 subscales by 12-40 percent points using the DACS device in combination with his conventional hearing aid in the contralateral ear.

In patient 02, there is a substantial improvement by 24-48 percent points in the EC, RV, BN subscales wearing the DACS system alone, when compared to the preoperative situation with a conventional hearing aid alone. Only the AV-scale shows a slight deterioration of 2%.

Similarly, patients 03 and 04 improve in all subscales, EC, RV and BN, by up to 29%, with disparate results in the AV-scale.

5.5 Discussion

The aim of this study was to demonstrate the proof of concept of the direct acoustic cochlear stimulator. It was shown that it is possible to develop and implant such a device. In our study, hearing and speech understanding was improved substantially in 4 patients with severe to profound mixed hearing losses.

Inner ear function did not deteriorate in any of our 4 patients. Nevertheless, the surgical procedure may be expected to have a similar risk of deafness as does stapedectomy [23, 24] or cochlear implantation [25, 26]. In our patients, hearing thresholds were in an order of magnitude that approached those of cochlear implant candidates. Therefore, we believe that small risk of deafness might be justifiable in light of the expected improvement, as it is also viewed as justified in stapedectomy, where the hearing of ears with much better thresholds is at stake.

The newly developed retromeatal approach has the advantage of being a minor and relatively fast surgical procedure. The most time consuming part of the implantation is the adaptation of the fixation system. The precise adaptation of the fixation system is important. However, further improvements in the fixation system design are possible and can further reduce implantation time. Several possibilities are currently being tested.

No problems or complications related to the transducer, the fixation system, the stapedectomy or the surgical procedure were observed in our patients. However, during the first two years postoperatively, two patients suffered from minor infections around the percutaneous plug of the investigational system. The device design is currently being modified. Among other improvements, the percutaneous plug will be replaced by a transcutaneous radio-frequency transmission similar to cochlear implants.

The conventional stapedectomy was successful in all of our patients with improvements between 14 and 28 dB (PTA). This is consistent with our own experience of more than 1500 stapedectomies at Inselspital, University of Berne, with an average improvement of 22 dB [23]. These improvements are reached already without the DACS being activated - a major difference compared to other implantable hearing systems, where patients may expect either unchanged thresholds or even some deterioration.

The DACS is a device with two acoustical inputs to the inner ear, namely through the DACS driven stapes prostheses and the tympano-ossicular chain through the conventional stapedectomy prosthesis. Our data show that there is no substantial interference between the two sound paths (comparison of the results with the ipsilaterally external auditory canal plugged and open, table 5.2 and 5.3). This is an expected result, as the signals differ by several orders of magnitude when the DACS is activated. In principle, the DACS can be implanted and would also work without the second stapes prosthesis.

The benefit of the DACS was measured using several methods: sound field hearing thresholds, speech intelligibility in quiet and in noise, as well as the assessment of subjective impressions using the APHAB questionnaire. A very encouraging result of our study is that all patients who participated in this study show better results in all of the above tests with the DACS than either preoperatively or when the DACS is switched off post-operatively. Our tests were taken 2 years after implantation, which suggests a stable long-term benefit. One of the most striking improvements in our data is the large increase in monosyllabic word recognition scores between 60 and 90 dB (Table 5.3). Differences of 40 to as much as 100 percent points at 75 dB SPL indicate a substantial benefit in everyday life. Besides speech understanding in quiet, there is also a substantial improvement in noise.

Comparisons with other implantable hearing aids are difficult, as our group of patients suffers from considerably higher hearing loss with (78-101 dB PTA) than published for other devices such as the fully implantable ossicular stimulator (MET) with 40-80 dB PTA [27] or the floating mass transducer (FMT) at the round window (65-85 dB PTA [3]). Generally, the gain in terms of speech understanding and improved hearing threshold seems to be higher than reported for other implantable hearing aids, especially in the lower frequency range. A conclusive comparison is beyond the scope of this first report.

One of the results of our investigation is that current fitting algorithms for conventional hearing aids can be applied to the DACS, but they must be modified sensibly for best results. Neither the AC-thresholds nor the BC-thresholds alone offer a reliable base for initial fittings. In our limited experience, measurements of hearing threshold via the DACS yielded the best initial fits. However, this is still an area to be explored in further research.

Further work is also planned and required in other areas related to the DACS. An improved device featuring transcutaneous transmission is currently in test. Furthermore, other coupling sites, e.g. the round window or the mobile footplate, will be considered. Eventually, other groups of patients, e.g. patients with radical cavities, difficult or failed tympanoplasties, possibly even patients with pure sensorineural hearing loss, may be considered for implantation. In such cases, a single stapes prosthesis connected to the DACS would be inserted through a small stapedotomy perforation, in order to further reduce the risk of an inner ear damage. The long-term goal is the development of a totally implantable system.

In summary, we presented data that show that implantation of the presented DACS system is a useful and efficient therapy for patients with severe to profound mixed hearing loss due to otosclerosis.

5.6 References

[1] Chen DA, Backous DD, Arriaga MA, Garvin R, Kobylek D, Littman T, Walgren S, Lura D: Phase 1 clinical trial results of the Envoy System: a totally

implantable middle ear device for sensorineural hearing loss. Otolaryngol Head Neck Surg 2004;131:904-16.

[2] Jorge JR, Pfister M, Zenner HP, Zalaman IM, Maassen MM: In vitro model for intraoperative adjustments in an implantable hearing aid (MET). Laryngoscope 2006;116:473-81.

[3] Colletti V, Soli SD, Carner M, Colletti L: Treatment of mixed hearing losses via implantation of a vibratory transducer on the round window. Int J Audiol 2006;45:600-8.

[4] Kasic JF, Fredrickson JM: The Otologics MET ossicular stimulator. Otolaryngol Clin North Am 2001;34:501-13.

[5] Ko WH, Zhu WL, Kane M, Maniglia AJ: Engineering principles applied to implantable otologic devices. Otolaryngol Clin North Am 2001;34:299-314.

[6] Zenner HP, Leysieffer H: [Active electronic cochlear implants for middle and inner ear hearing loss--a new era in ear surgery. I: Basic principles and recommendations on nomenclature]. Hno 1997;45:749-57.

[7] Huttenbrink KB: Current status and critical reflections on implantable hearing aids. Am J Otol 1999;20:409-15.

[8] Ball: Implantable and external hearing systems having a floating mass transducer. 772779, 1999.

[9] Needham AJ, Jiang D, Bibas A, Jeronimidis G, O'Connor AF: The effects of mass loading the ossicles with a floating mass transducer on middle ear transfer function. Otol Neurotol 2005;26:218-24.

[10] Parkin JL, Parkin MJ: Multichannel cochlear implantation with percutaneous pedestal. Ear Nose Throat J 1994;73:156-8, 163-4.

[11] Heiland KE, Goode RL, Asai M, Huber AM: A human temporal bone study of stapes footplate movement. Am J Otol 1999;20:81-6.

[12] Stieger C, Bernhard H, Waeckerlin D, Kompis M, Burger J, Haeusler R: Human temporal bones versus a mechanical model to evaluate three middle-ear transducers. J Rehabil Res Dev in press.

[13] Voss SE, Rosowski JJ, Merchant SN, Peake WT: Acoustic responses of the human middle ear. Hear Res 2000;150:43-69.

[14] Bernhard H, Stieger C, Perriard Y: New implantable hearing device based on a micro-actuator that is directly coupled to the inner ear fluid, EMBC06, 2006b, p IEEE Engineering in Medicine and Biology Society Proceeding.

[15] Bernhard H, Fontannaz J, Peclat C, Haller M, Cauwels K, Kloeck B, Huybrechts K, Stieger C, Haeusler R, Kaiser T: Implantable Actuator for Hearing aid Applications. 2006a.

[16] Hausler R: Cochlear implantation without mastoidectomy: the pericanal electrode insertion technique. Acta Otolaryngol 2002;122:715-9.

[17] Kompis M: Sprachaudiometrie; M. Kompis, (ed): Audiologie. Bern, Hans Huber, 2004a.

[18] Tschopp K, Zust H: Performance of normally hearing and hearing-impaired listeners using a German version of the SPIN test. Scand Audiol 1994;23:241-7.

[19] Todt I, Seidl RO, Ernst A: Hearing benefit of patients after Vibrant Soundbridge implantation. ORL J Otorhinolaryngol Relat Spec 2005;67:203-6.

[20] Cox RM, Alexander GC: The abbreviated profile of hearing aid benefit. Ear Hear 1995;16:176-86.

[21] Kompis M: Überhören und Vertäuben; M. Kompis, (ed): Audiologie. Bern, Hans Huber, 2004b.

[22] Kompis M, Krebs M, Haeusler R: Speech understanding in quiet and in noise with bone-anchored hearing aids BAHA compact and BAHA Divino. Acta Otolaryngol 2007;127:829-835.

[23] Haeusler R: Fortschritte in der Stapeschirurgie. Laryngo-Rhino-Otol 2000;79 Supplement 2:95–139.

[24] Shea JJ, Jr.: Forty years of stapes surgery. Am J Otol 1998;19:52-5.

[25] Dutt SN, Ray J, Hadjihannas E, Cooper H, Donaldson I, Proops DW: Medical and surgical complications of the second 100 adult cochlear implant patients in Birmingham. J Laryngol Otol 2005;119:759-64.

[26] Green KM, Bhatt YM, Saeed SR, Ramsden RT: Complications following adult cochlear implantation: experience in Manchester. J Laryngol Otol 2004;118:417-20.

[27] Jenkins HA, Atkins JS, Horlbeck D, et al.: U.S. Phase I preliminary results of use of the Otologics MET Fully-Implantable Ossicular Stimulator. Otolaryngol Head Neck Surg 2007;137:206-212.

Chapter 6: Conclusion and Outlook

This chapter includes a general conclusion for both concepts which have been developed and evaluated during this thesis. An outlook is provided further on.

In this thesis, two different kinds of implantable hearing systems have been evaluated and developed. They work on different principles and focus on different indications. The CLT (Contactless Transducer) drives the ossicular chain and is intended for the therapy of patients with purely sensorineural hearing loss. The DACS (Direct Acoustical Cochlear Stimulation) stimulates the inner ear fluid directly at the stage of the round window and is focused on patients with combined hearing loss.

For the geometrical design of both transducers, a data set describing quantitatively the adult human middle ear was needed and therefore generated. In principle, the method employed in this study using standard CT-scans can also be used preoperatively to rule out exclusion criteria.

6.1 Contactless transducer (CLT)

The CLT is based on a electromagnetic design with a coil and a permanent magnet. It has two major advantages. It is, in contrast other systems, minimally invasively implantable through the external auditory canal. The coil is fixed at the wall of the middle ear cavity and the magnet on the incus. Due to the contactless concept, it does not generate a static preload on the ossicular chain, as do other systems.

The CLT was optimized using computer simulations for different coil designs and for a range of radial displacements and air gaps. A subset of the simulation results is verified experimentally in a laboratory setting. Results from the simulations and the experiments were found to be in reasonable agreement. It was shown that the proposed transducer design could, at a size and geometry which should allow an implantation through the external auditory canal, provide an acoustic output corresponding to 120 dB SPL. The CLT can be optimized either to maximize output levels or to be tolerant of radial displacements of up to 1 mm between coil and magnet. Based on this study and the morphometric data, a tolerant CLT was designed.

The CLT was evaluated using the life-size mechanical middle ear model. Compared to tests using human temporal bones, the handling was found to be simpler and less time consuming. Furthermore, individual parameters could be varied systematically more easily than with human temporal bones. Results from the mechanical middle ear model and from human temporal bones were found to be in reasonable agreement with all three transducer designs considered in this research. Regarding systematic variations of the mounting parameters, the CLT system was found to become non-linear for large offsets and contact between the components. The output is efficient for low and high frequencies. However, in the important frequencies of the speech field, the generated output is insufficient, also in comparison to other implantable middle ear transducers (FMT, DRT) and the conventional hearing aid speakers. The design exhausts the geometry of the middle ear cavity and the accessible power. Therefore, an improvement in the output in the mentioned frequencies is hardly possible.

6.2 Direct acoustical cochlear stimulation (DACS)

The second transducer was developed for patients with severe conductive hearing loss, for which no efficient therapy is available at the present time. This implantable hearing system works on the principle of direct acoustical cochlear stimulation (DACS) which is in contrast to middle ear transducers. The DACS transducer drives a conventional stapes prosthesis coupled to the liquid of the inner ear. The DACS system was implanted first in temporal bones and then in isolated human heads and showed an equivalent sound pressure levels of 140 dB SBL (125 dB broadband) applied to the inner ear fluid. The transducer was fixed in the bone behind the ear. A percutaneous plug provides the interface to the externally worn audio-processor. The surgical intervention is a combination of cochlear implantation and stapedectomy and must consequently be performed by an experienced otologist.

Experience in the first clinical studies show that patients with severe combined hearing loss will accept an implantation that is not categorically minimally invasive, as they have a strong psychological strain induced by their hearing loss.

The device was for the first time implanted in three patients in a pilot clinical study at the ENT Department, Inselspital, University of Berne, Switzerland. A fourth patient was implanted at the MHH (Medizinische Hochschule Hannover, Germany). All patients suffered a severe combined otosclerotic hearing loss. Postoperative audiological results with three implanted patients are outstanding. They show that patients with severe combined otosclerotic hearing loss reach substantially higher hearing improvement with the DACS than is obtained by otological surgery or conventional hearing aids alone.

6.3 Outlook

Implantable middle ear transducers such as the CLT are in direct competition with conventional hearing systems. Ideally, a middle ear transducer should be able to perform significantly better than conventional hearing aids and be implantable with a minimal surgical intervention. Unless both criteria are met, presumably they will hardly find wide acceptance by audiologists and patients. They will endure with patients who have medical contraindications or strong aversion towards conventional hearing aids. As the CLT concept was not able to fulfil both conditions, it has not been propagated.

The DACS, on the other hand, has been developed further. The results of the clinical trail were outstanding. The pilot study showed high efficiency with the DACS in patients with severe combined hearing loss due to otosclerosis. However, further developments are necessary and currently under development. Second generation of DACS devices will provide trancutaneous inductive transmission instead of the percutaneous plug. A newly developed fixation system will simplify the surgical procedure. A specially developed fitting strategy is being developed to ameliorate

Conclusion

programming of the external audio processor. An ongoing PhD thesis (H. Bernhard) elaborates on technology for further development of the DACS transducer.

With this second generation, a multicenter study in Europe and USA will start in the year 2009 to expand the indications for the DACS device.

Die VDM Verlagsservicegesellschaft sucht für wissenschaftliche Verlage abgeschlossene und herausragende

Dissertationen, Habilitationen, Diplomarbeiten, Master Theses, Magisterarbeiten usw.

für die kostenlose Publikation als Fachbuch.

Sie verfügen über eine Arbeit, die hohen inhaltlichen und formalen Ansprüchen genügt, und haben Interesse an einer honorarvergüteten Publikation?

Dann senden Sie bitte erste Informationen über sich und Ihre Arbeit per Email an *info@vdm-vsg.de*.

Sie erhalten kurzfristig unser Feedback!

VDM Verlagsservicegesellschaft mbH
Dudweiler Landstr. 99   Telefon +49 681 3720 174
D - 66123 Saarbrücken   Fax   +49 681 3720 1749
www.vdm-vsg.de

Die VDM Verlagsservicegesellschaft mbH vertritt

Printed by Books on Demand GmbH, Norderstedt / Germany